Norman Macleod Ferrers

An Elementary Treatise on Trilinear Co-Ordinates

The method of reciprocal polars, and the theory of projections

Norman Macleod Ferrers

An Elementary Treatise on Trilinear Co-Ordinates
The method of reciprocal polars, and the theory of projections

ISBN/EAN: 9783337322946

Printed in Europe, USA, Canada, Australia, Japan

Cover: Foto ©berggeist007 / pixelio.de

More available books at **www.hansebooks.com**

AN ELEMENTARY TREATISE

ON

TRILINEAR CO-ORDINATES,

THE METHOD OF RECIPROCAL POLARS, AND THE THEORY OF PROJECTIONS.

BY THE REV.

N. M. FERRERS, D.D., F.R.S.

MASTER OF GONVILLE AND CAIUS COLLEGE,
HON. LL.D. OF THE UNIVERSITY OF GLASGOW.

FOURTH EDITION.

London:
MACMILLAN AND CO.
AND NEW YORK.
1890

[*All Rights reserved.*]

PREFACE.

THE prominence which the modern geometrical methods have recently acquired in the studies of the University of Cambridge, appears to justify the publication of a treatise devoted exclusively to these branches of Mathematics. This remark applies more especially to the method of Trilinear Co-ordinates, which forms the subject of the greater part of the following work. My object in writing on this subject has mainly been to place it on a basis altogether independent of the ordinary Cartesian system, instead of regarding it as only a special form of Abridged Notation.

A desire not unduly to increase the size of the book has prevented me from proceeding beyond Curves of the Second Degree.

In this Second Edition several new articles have been added, especially in the latter part of the work, and the chapter on Reciprocal Polars considerably enlarged.

N. M. F.

GONVILLE AND CAIUS COLLEGE,
August, 1866.

In the Third Edition, I have rewritten some articles where the demonstrations were imperfect or obscure, and have added some examples, taken from various Cambridge Examination papers.

DECEMBER, 1875.

CONTENTS.

CHAPTER I.

TRILINEAR CO-ORDINATES. EQUATION OF A STRAIGHT LINE.

ARTS.		PAGE
1.	DEFINITION of Trilinear Co-ordinates	1
2.	Identical relation between the Trilinear Co-ordinates of a Point	ib.
3.	Distance between two given Points	4
4—6.	Investigation of Equations of certain Straight Lines . . .	6
7.	Every Straight Line may be represented by an Equation of the First Degree	9
8.	Every Equation of the First Degree represents a Straight Line .	11
9.	Point of Intersection of Two Straight Lines	12
10.	Equation of a Straight Line passing through Two given Points .	ib.
11.	Equation of a Straight Line passing through the Point of Intersection of Two given Straight Lines	13
12.	Condition that Three Points may lie in the same Straight Line .	ib.
13.	Condition that Three Straight Lines may intersect in a Point .	14
14.	Condition that Two Straight Lines may be parallel to one another. Line at Infinity	ib.
15.	Equation of a Straight Line, drawn through a given Point, parallel to a given Straight Line	17
16.	Inclination of a Straight Line to a side of the Triangle of Reference	18
17.	Condition of Perpendicularity	19
18.	Distance from a Point to a Straight Line	20
	EXAMPLES	21
19.	ANHARMONIC RATIO. Definitions	23
20.	The Anharmonic Ratio of a Pencil is equal to that of the range in which it is cut by any Transversal	ib.
21.	Definition of an Harmonic Pencil	25

viii   CONTENTS.

ARTS.		PAGE
22.	The Bisectors of any Angle form, with the Lines containing it, an Harmonic Pencil	25
23.	Anharmonic Ratio of a given Pencil	26
24.	Fourth Harmonic to Three given Straight Lines . . .	27
25.	Harmonic Relation of Points and Lines	28
26.	ON INVOLUTION. Definitions	30
27—29.	Anharmonic Properties of Points and Lines in Involution .	31

CHAPTER II.

SPECIAL FORMS OF THE EQUATION OF THE SECOND DEGREE.

1.	Every Equation of the Second Degree represents a Conic Section	33
2, 3.	Equation of the Conic described about the Triangle of Reference	34
4.	Position of the Centre. Condition for a Parabola . . .	35
5.	Condition of Tangency. Every Parabola touches the Line at Infinity	37
6.	Equation of the Circumscribing Circle	ib.
7.	Equation of the Conic touching the Three Sides of the Triangle of Reference	39
8.	Position of the Centre. Condition for a Parabola . . .	40
9.	Condition of Tangency	42
10.	Equations of the Four Circles which touch the Three Sides of the Triangle of Reference	43
11—15.	Equation involving the Squares only of the Variables .	45
16.	Condition of Tangency	48
17.	Condition for a Parabola	49
18.	Co-ordinates of the Centre.	ib.
19.	Equation of the Circle, with respect to which the Triangle of Reference is self-conjugate	51
20.	Equation of the Conic which touches two sides of the Triangle of Reference in the points where they meet the third . .	52
21.	Any Chord of a conic is divided *harmonically* by the Conic, any Point, and its Polar	53
23.	Equation of a Line joining Two given Points . . .	ib.
24.	Equation of the Tangent at a given Point	54
25.	Pole of a given Straight Line	ib.
26.	Condition of Tangency. Condition for a Parabola . .	ib.
27.	Co-ordinates of the Centre.	55
	EXAMPLES	56

CHAPTER III.

ELIMINATION BETWEEN LINEAR EQUATIONS.

ARTS.		PAGE
2.	Definition of a Determinant	59
3—6.	Law of Formation of Determinants	60
7.	Signs of the several Terms of a Determinant . . .	65
8.	Sign changed by interchange of Two Consecutive Lines or Columns	66
9.	Multiplication of a Determinant by a given Quantity . .	67
10.	Minors of a Determinant	68
11.	Condition that a Quadratic Function may be resolvable into Two Factors	*ib.*
12, 13.	Pascal's Theorem	69
	EXAMPLES	71

CHAPTER IV.

ON THE CONIC REPRESENTED BY THE GENERAL EQUATION OF THE SECOND DEGREE.

2.	To find the point in which a straight line, drawn in a given direction through a given point of the conic, meets the conic again	74
3.	Equation of the Tangent at a given Point	75
4, 5.	Condition that a given Straight Line may touch the Conic	76
6.	Condition for a Parabola	78
7.	Condition that the Conic may break up into Two Straight Lines	*ib.*
8.	Equation of the Polar of a given Point	*ib.*
9.	Co-ordinates of the Pole of a given Straight Line . .	79
10.	Equation of the pair of Tangents drawn to the Conic from a given external Point	80
11.	Co-ordinates of the Centre	82
12.	Equation of the Asymptotes	83
13.	Condition for a Rectangular Hyperbola	84
14.	Conditions for a Circle	86

ARTS.		PAGE
15.	All Circles pass through the same two points at infinity	88
16.	All Conics, similar and similarly situated to each other, intersect in the same two points in the line at infinity	89
17.	Radical axis of two similar and similarly situated Conics	ib.
18.	Property of the nine-point Circle	91
19.	Equation of the nine-point Circle	92
20.	Locus of the intersection of two Tangents at right angles to one another. Directrix of a Parabola	93
21.	To find the magnitudes of the axes of the Conic	94
22.	To find the area of the Conic. Criterion to distinguish between an Ellipse and an Hyperbola	96
	EXAMPLES	98

CHAPTER V.

TRIANGULAR CO-ORDINATES.

1.	Definition of the Triangular Co-ordinates of a Point	100
2.	Formulæ relating to Straight Lines	ib.
3.	Formulæ relating to Conics	102

CHAPTER VI.

RECIPROCAL POLARS.

3.	Definition of a Polar Reciprocal	105
5.	The degree of a curve is the same as the class of its reciprocal, and *vice versâ*	106
6, 7.	The polar reciprocal of a conic is a conic	107
8.	Equation of the Polar Reciprocal of one Conic with regard to another	ib.
10.	Instances of Transformation	109
12.	*Brianchon's Theorem*	112
13.	The anharmonic ratio of the Pencil, formed by four intersecting straight lines, is the same as that of the range formed by their poles	113
15.	Any straight line drawn through a given point A is divided harmonically by any Conic Section and the polar of A with respect to it	115

CONTENTS.

ARTS.		PAGE
17.	If four straight lines form an harmonic pencil, either pair will be its own polar reciprocal with respect to the other	116
18.	Condition that two pairs of straight lines may form an harmonic pencil	ib.
20.	Reciprocation with respect to a Circle	118
23.	Reciprocation with respect to a Point	119
24.	The three circles described on the diagonals of a complete quadrilateral as diameters have a common radical axis	ib.
	Foci of a quadrilateral	ib.
25.	Orthocentre of a triangle	ib.
28.	The director circles of all conics which touch four given straight lines have a common radical axis	121
29.	Polar reciprocal of a circle with regard to any point	ib.
30.	Instances of Transformation of Theorems by Reciprocation with respect to a point	123
31.	Corresponding Points and Lines. The angle between the radius vector and tangent in any curve is equal to the corresponding angle in the Reciprocal Curve	125
33.	Co-ordinates of the foci of a Conic	126
34.	Double application of the Method of Reciprocal Polars	128
	EXAMPLES	ib.

CHAPTER VII.

TANGENTIAL CO-ORDINATES.

1.	Definition of the Tangential Co-ordinates of a Straight Line	131
2.	Interpretation of the Negative Sign. Equations of certain points	ib.
3.	General Equation of a Point	133
4.	Identical relation between the co-ordinates of any straight Line	134
6.	An equation of the nth degree represents a curve of the nth class	136
7.	Equation of a Conic, touching the three sides of the triangle of reference	ib.
8.	Equation of circumscribed Conic	138
9.	Equation of the Pole of a given straight line, and of the centre. Condition for a Parabola	139
10.	Circular points at infinity	ib.
	Conditions for a Circle	140
11.	Self-conjugate Conic	141

ARTS.		PAGE
	EXAMPLES	141
12.	Tangential rectangular Co-ordinates	142
15.	Tangential polar Co-ordinates	144
	EXAMPLES	145

CHAPTER VIII.

ON THE INTERSECTION OF CONICS, AND ON PROJECTIONS.

1—3.	Any two conics intersect in four points, real or imaginary. Vertices of the quadrangle formed by these points . .	146
4—7.	If the four points of intersection be all real, or all imaginary, all the vertices are real. If two of the points of intersection be real, and two imaginary, one vertex only is real. If the four points of intersection be all real, all the common chords are real; if not, one pair only is real . . .	147
8.	Invariants of two Conics	150

ON PROJECTIONS.

9.	Definition of Projections	152
10.	Projection to infinity	ib.
13.	Any quadrilateral may be projected, in an infinite number of ways, into a parallelogram, of which the angles are of any magnitude	153
14.	Projection of Tangents, of Poles, and Polars	ib.
15.	Any two conics may be projected into concentric curves .	154
16.	Also into similar and similarly situated curves . . .	ib.
17.	These projections may be effected in an infinite number of ways	ib.
18.	Any two intersecting conics may be projected into hyperbolas of any assigned eccentricity	ib.
19.	*Any two conics may be projected into conics of any eccentricity, or into circles*	155
20.	Projection of the foci and directrices of a Conic . . .	ib.
21.	The anharmonic ratio of any pencil or range is unaltered by projection	156
22.	Any two lines, which make an angle A with each other, form with the lines joining the circular points at infinity to their point of intersection, a pencil of which the anharmonic ratio is $e^{(\pi-2A)\sqrt{-1}}$	157

CONTENTS.

ARTS.		PAGE
23.	The anharmonic ratio of any four points on, or any four tangents to, a conic, is constant	158
25.	Any system of points in involution projects into a system in involution, and the foci of one system project into the foci of the other	ib.
26.	A system of Conics, passing through four given points, cut any straight line in a system of points in involution . . .	159
27.	Orthogonal Projection	ib.
	EXAMPLES	160

CHAPTER IX.

MISCELLANEOUS PROPOSITIONS.

ON THE DETERMINATION OF A CONIC FROM FIVE GIVEN GEOMETRICAL CONDITIONS.

2.	If five points be given, one conic only can be drawn . .	162
3.	If four points and one tangent be given, two conics can be drawn	ib.
4.	If three points and two tangents be given, four conics can be drawn	163
5.	If two points and three tangents be given, four conics can be drawn	ib.
6.	If one point and four tangents be given, two conics can be drawn	164
7.	If five tangents be given, one conic only can be drawn . .	ib.
8.	Reduction of certain other conditions to these . . .	ib.
9.	Conjugate triad	165

ON THE LOCUS OF THE CENTRE OF A SYSTEM OF CONICS WHICH SATISFY FOUR CONDITIONS.

11.	Four points given	165
12.	Three points and a tangent	167
13.	Two points and two tangents	168
14.	One point and three tangents	169
15.	Four tangents	171

SUPPLEMENTARY PROBLEMS.

ARTS.		PAGE
16.	The product of any two determinants is a determinant	171
17.	Property of the co-ordinates of three points, forming a conjugate triad	173
18.	Envelope of a side of an inscribed triangle whose other sides pass each through a fixed point	174
19.	Locus of a vertex of a circumscribed triangle, whose other vertices move each along a fixed straight line	176
20.	Trilinear co-ordinates of the Foci	177
	MISCELLANEOUS EXAMPLES	179

TRILINEAR CO-ORDINATES.

CHAPTER I.

TRILINEAR CO-ORDINATES. EQUATION OF A STRAIGHT LINE.

1. IN the system of co-ordinates ordinarily used, the position of a point in a plane is determined by means of its distances from two given straight lines. In the system of which we are about to treat, the position of a point in a plane will be determined by the ratios of its distances from three given straight lines in that plane, these straight lines not passing through the same point. The triangle formed by these three straight lines is called *the triangle of reference*, its sides, *lines of reference*, and the distances of a point from its three sides will be called *the trilinear co-ordinates* of that point. We shall usually denote the angular points of the triangle of reference by the letters A, B, C, the lengths of the sides respectively opposite to them by a, b, c, and the distances of any point from BC, CA, AB respectively by the letters α, β, γ.

When two points lie on opposite sides of a line of reference, the distance of one of these points from that line may be considered as positive, and that of the other as negative. We shall consider α, the distance of a point from the line BC, as positive if the point lie on the same side of that line as the point A does, negative if on the other side; and similarly for β and γ. It thus appears that the trilinear co-ordinates of any point within the triangle of reference are all positive; while no point has all its co-ordinates negative.

2. Between the trilinear co-ordinates of any point an important relation exists, which we proceed to investigate.

If Δ denote the area of the triangle of reference, α, β, γ, the trilinear co-ordinates of any point, then

$$a\alpha + b\beta + c\gamma = 2\Delta.$$

Let P be the given point, and first suppose it to lie within the triangle of reference (fig. 1). Join PA, PB, PC, and draw PD perpendicular to BC. Then $PD = \alpha$, and $a\alpha =$ twice the area of the triangle PBC.

Fig. 1.

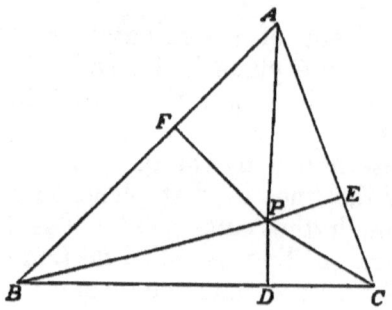

Similarly $b\beta =$ twice the area of PCA,
$c\gamma =$ twice the area of PAB.

Adding these equations, we get
$$a\alpha + b\beta + c\gamma = 2\Delta.$$

Next, suppose P to lie between AB, AC produced, and on the side of BC remote from A (fig. 2). Then α will be

Fig. 2.

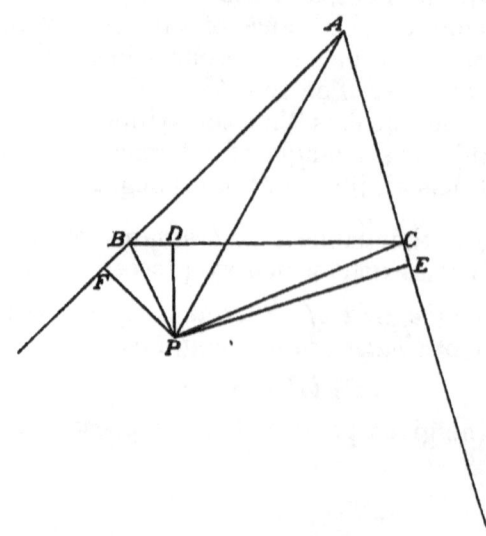

negative, while β, γ are positive. Hence, twice the area PBC will be represented by $-a\alpha$, and we shall therefore have as before

$$a\alpha + b\beta + c\gamma = 2\Delta.$$

Thirdly, let P lie between AB, AC, produced backwards (fig. 3), so that β, γ are negative while α is positive. Twice

Fig. 3.

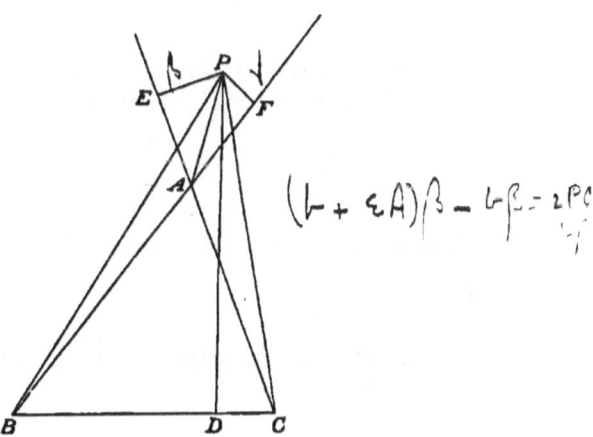

the areas of PBC, PCA, PAB, are now represented by $a\alpha$, $-b\beta$, $-c\gamma$ respectively, so that we still have

$$a\alpha + b\beta + c\gamma = 2\Delta.$$

In all cases, therefore,

$$a\alpha + b\beta + c\gamma = 2\Delta.$$

The importance of the above proposition arises from its enabling us to express any equation in a form homogeneous with respect to the trilinear co-ordinates of any point to which it relates. Any locus may be represented, as in the ordinary system, by means of a relation between two co-ordinates, β and γ for example, and this may be made homogeneous in α, β, γ by multiplying each term by $\dfrac{a\alpha + b\beta + c\gamma}{2\Delta}$, raised to a suitable power. Thus, the equation $\beta^2 + h\gamma + k^2 = 0$ is equivalent to the homogeneous equation

$$4\Delta^2\beta^2 + 2\Delta h\gamma(a\alpha + b\beta + c\gamma) + k^2(a\alpha + b\beta + c\gamma)^2 = 0.$$

The following examples may familiarize the reader with this system of co-ordinates.

1. Prove that the co-ordinates of the middle point of the line BC are $0, \dfrac{\Delta}{b}, \dfrac{\Delta}{c}$.

2. The co-ordinates of the centre of the circumscribed circle are $R\cos A$, $R\cos B$, $R\cos C$, where

$$R = \frac{2\Delta}{a\cos A + b\cos B + c\cos C}.$$

3. The co-ordinates of the centre of the inscribed circle are each equal to $\dfrac{2\Delta}{a+b+c}$.

What are the co-ordinates of the centres of the escribed circles?

4. The co-ordinates of the centre of gravity are

$$\frac{2\Delta}{3a}, \frac{2\Delta}{3b}, \frac{2\Delta}{3c}.$$

5. Prove that $a\sin A + \beta\sin B + \gamma\sin C$ is equal to $\dfrac{\Delta}{R}$; where R is the radius of the circumscribing circle.

3. *To find the distance between two given points, in terms of their trilinear co-ordinates.*

Let $\alpha_1, \beta_1, \gamma_1$; $\alpha_2, \beta_2, \gamma_2$, be the co-ordinates of two given points, r the distance between them.

Then, r^2 will be a rational integral function of $\alpha_1 - \alpha_2$, $\beta_1 - \beta_2$, $\gamma_1 - \gamma_2$, of the second degree *.

---

* This, if not self-evident, may be proved as follows:

Let P, Q be the two given points. Join PQ, and draw PM, QM' perpen-

DISTANCE BETWEEN TWO POINTS. 5

Again, since
$$a\alpha_1 + b\beta_1 + c\gamma_1 = 2\Delta,$$
$$a\alpha_2 + b\beta_2 + c\gamma_2 = 2\Delta;$$
$$\therefore a(\alpha_1 - \alpha_2) + b(\beta_1 - \beta_2) + c(\gamma_1 - \gamma_2) = 0;$$
$$\therefore (\alpha_1 - \alpha_2)^2 = -\frac{c}{a}(\gamma_1 - \gamma_2)(\alpha_1 - \alpha_2) - \frac{b}{a}(\alpha_1 - \alpha_2)(\beta_1 - \beta_2).$$

Similar expressions may be found for $(\beta_1 - \beta_2)^2$, $(\gamma_1 - \gamma_2)^2$.

Hence, r^2 will be of the form
$$l(\beta_1 - \beta_2)(\gamma_1 - \gamma_2) + m(\gamma_1 - \gamma_2)(\alpha_1 - \alpha_2) + n(\alpha_1 - \alpha_2)(\beta_1 - \beta_2),$$

dicular to AB, PN, QN' to AC. Draw Qm perpendicular to PM, Qn to PN, and join mn. Then
$$r = PQ = \frac{mn}{\sin mPn}$$
$$= \frac{mn}{\sin A},$$
and $Pn = \beta_1 - \beta_2$, $Pm = \gamma_1 - \gamma_2$;

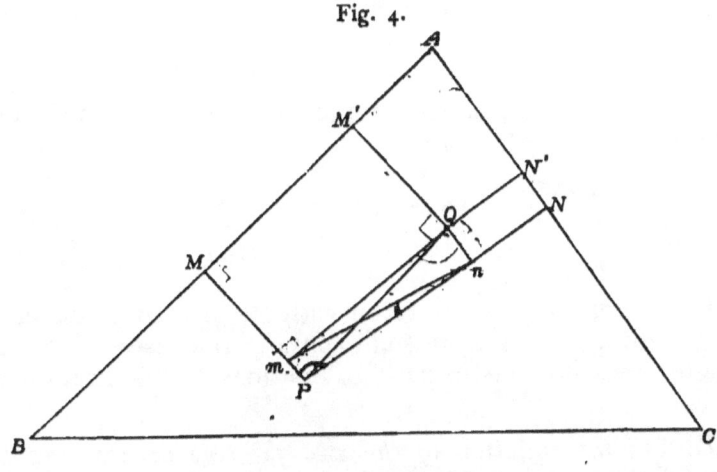

Fig. 4.

$$\therefore mn^2 = (\beta_1 - \beta_2)^2 + (\gamma_1 - \gamma_2)^2 + 2(\beta_1 - \beta_2)(\gamma_1 - \gamma_2)\cos A,$$
$$\text{whence } r^2 = \frac{(\beta_1 - \beta_2)^2 + (\gamma_1 - \gamma_2)^2 + 2(\beta_1 - \beta_2)(\gamma_1 - \gamma_2)\cos A}{\sin^2 A},$$

a rational integral function of the second degree.

where l, m, n are certain functions of a, b, c, which we proceed to determine.

Since the values of l, m, n are independent of the positions of the points, the distance of which we wish to find, suppose these points to be B and C. Then

$$\alpha_1 = 0, \ \beta_1 = \frac{2\Delta}{b}, \ \gamma_1 = 0,$$

$$\alpha_2 = 0, \ \beta_2 = 0, \ \gamma_2 = \frac{2\Delta}{c};$$

also $r = a$. Hence

$$a^2 = -l\frac{2\Delta}{b} \cdot \frac{2\Delta}{c};$$

$$\therefore l = -\frac{a^3bc}{4\Delta^2}.$$

Similarly $$m = -\frac{ab^3c}{4\Delta^2},$$

$$n = -\frac{abc^3}{4\Delta^2}.$$

Hence $r^2 = -\dfrac{abc}{4\Delta^2} \{a(\beta_1 - \beta_2)(\gamma_1 - \gamma_2) + b(\gamma_1 - \gamma_2)(\alpha_1 - \alpha_2)$
$$+ c(\alpha_1 - \alpha_2)(\beta_1 - \beta_2)\}.$$

This is one form of the expression for r^2. It may also be proved in a similar manner that

$$r^2 = \frac{abc}{4\Delta^2} \{a \cos A (\alpha_1 - \alpha_2)^2 + b \cos B (\beta_1 - \beta_2)^2$$
$$+ c \cos C (\gamma_1 - \gamma_2)^2\}.$$

4. We next proceed to investigate the equation of a straight line; and first, we shall consider the cases of certain straight lines bearing important relations to the triangle of reference.

To find the equation of the straight line drawn through one of the angular points of the triangle of reference, so as to bisect the opposite side.

Let D be the middle point of the side BC, we have then to investigate the equation of the straight line AD.

EQUATIONS OF THE BISECTORS.

In AD take any point P, and let α, β, γ be its co-ordi-

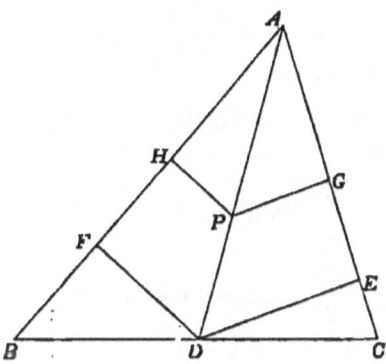

Fig. 5.

nates. From D, P draw DE, PG perpendicular to AC, DF, PH perpendicular to AB. Then by similar triangles

$$PG : DE :: PH : DF.$$

But $$DE \cdot AC = DF \cdot AB,$$

for each is equal to the area of the triangle ABC.

Hence $$PG \cdot AC = PH \cdot AB,$$

or $$b\beta = c\gamma.$$

This is a relation between the co-ordinates of any point on the line AD, it therefore is the equation of that line.

Cor. It hence may be proved that the three straight lines, drawn through the angular points of a triangle to bisect the opposite sides, intersect in a point. For these straight lines will be represented by the equations

$$b\beta = c\gamma,$$
$$c\gamma = a\alpha,$$
$$a\alpha = \beta b,$$

and, therefore, all pass through the point for which

$$a\alpha = b\beta = c\gamma.$$

In the next three propositions the reader will easily be able to draw a figure for himself, by comparison with fig. 5.

5. *To find the equation of the straight line drawn through one of the angular points of the triangle of reference, perpendicular to the opposite side.*

Making a construction similar to that in the last proposition, it will be seen that we have here

$$\frac{DE}{AD} = \sin DAE = \cos C,$$

$$\frac{DF}{AD} = \sin DAF = \cos B;$$

$$\therefore \frac{DE}{\cos C} = \frac{DF}{\cos B}.$$

Hence $\quad \dfrac{PG}{\cos C} = \dfrac{PH}{\cos B};$

$$\therefore PG \cos B = PH \cos C,$$

or $\quad \beta \cos B = \gamma \cos C.$

This will be the equation of the straight line, drawn through A, at right angles to BC.

COR. It may hence be shewn that the three straight lines drawn through the angular points of a triangle, perpendicular to the opposite sides, intersect in the point determined by the equations

$$\alpha \cos A = \beta \cos B = \gamma \cos C.$$

6. *To find the equations of the internal and external bisectors of an angle of the triangle of reference.*

For the internal bisector of the angle A, we shall have, making the same construction as before,

$$PG = PH.$$

The straight line will be therefore represented by the equation $\beta = \gamma$.

For the external bisector we proceed as follows. Let Q be any point on the line, α, β, γ its co-ordinates. Draw QK perpendicular to AC, QL to AB. Then, as before, we have

$$QK = QL.$$

It will however be observed, that if Q and B lie on the same side of AC, Q and C will lie on opposite sides of AB, and *vice versâ*. Hence, if

$$QK = \beta, \quad QL = -\gamma.$$

We have therefore

$$\beta + \gamma = 0$$

as the equation of the line AQ, which externally bisects the angle A.

From the form of these equations we see, (1), That the three internal bisectors of the angles of a triangle intersect in a point; (2), That the internal bisector of any one angle, and the external bisectors of the other two, also intersect in a point.

These points may be shewn to be respectively the centres of the inscribed and escribed circles.

We shall hereafter prove that the points, in which the external bisectors of each angle respectively intersect the sides opposite to them, lie in the same straight line; and also that the points in which the external bisector of any one angle, and the internal bisectors of the other two angles, intersect the sides respectively opposite to them, lie in the same straight line.

7. We now proceed to investigate the general equation of a straight line.

Every straight line may be represented by an equation of the first degree.

Let Q be any point on the straight line AC, R on AB, and P any point on the straight line QR, then we have to investigate the relation between the co-ordinates (α, β, γ) of the point P.

Fig. 6.

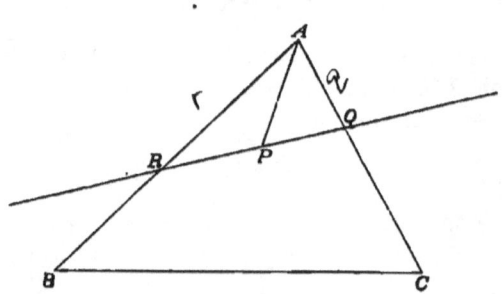

The property of the straight line, which we shall make the basis of our investigation, is, that it is the locus of a point which moves in such a manner, that the sum of the areas of the triangles PAQ, PAR is constant.

Let $AQ = q$, $AR = r$, then the areas of the triangles PAQ, PAR will be respectively represented by $\tfrac{1}{2}q\beta$, $\tfrac{1}{2}r\gamma$, and the area of QAR by $\dfrac{qr}{bc}\Delta$.

Hence
$$q\beta + r\gamma = \frac{2qr}{bc}\Delta$$
$$= \frac{qr}{bc}(a\alpha + b\beta + c\gamma).$$

This is the equation of the straight line QR, and, since it involves the two arbitrary quantities q, r, it is in the most general form of the equation of the first degree between two variables. Putting

$$\frac{qra}{bc} = l, \quad \frac{qr}{c} - q = m, \quad \frac{qr}{b} - r = n,$$

the equation may be written

$$l\alpha + m\beta + n\gamma = 0.$$

EQUATION OF THE FIRST DEGREE.

8. We shall next establish the converse proposition, that *every equation of the first degree represents a straight line.*

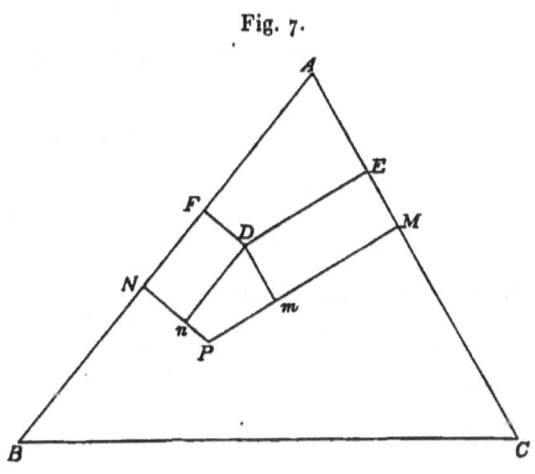

Fig. 7.

Let
$$l\alpha + m\beta + n\gamma = 0$$
be the general equation of the first degree, and let f, g, h be the co-ordinates of any fixed point D on the locus of the equation, α, β, γ those of any point P.

Draw DE, PM perpendicular to AC, DF, PN perpendicular to AB. Also draw Dm, Dn, perpendicular respectively to PM, PN.

Then $Pm = \beta - g$, $Pn = \gamma - h$.

Also, since f, g, h is a point on the locus,
$$lf + mg + nh = 0,$$
whence $\quad l(\alpha - f) + m(\beta - g) + n(\gamma - h) = 0.$

Again, $\quad a\alpha + b\beta + c\gamma = 2\Delta,$
$$af + bg + ch = 2\Delta;$$
$$\therefore a(\alpha - f) + b(\beta - g) + c(\gamma - h) = 0;$$
$$\therefore \frac{\alpha - f}{bn - cm} = \frac{\beta - g}{cl - an} = \frac{\gamma - h}{am - bl}.$$

Hence, the ratio of Pm to Pn is constant, whatever point on the locus P may represent. This can only be true when that locus is a straight line.

9. *To find the co-ordinates of the point of intersection of two given straight lines.*

Let the equations of the two straight lines be
$$l\alpha + m\beta + n\gamma = 0,$$
$$l'\alpha + m'\beta + n'\gamma = 0.$$

Where these intersect, we have
$$\frac{\alpha}{mn' - m'n} = \frac{\beta}{nl' - n'l} = \frac{\gamma}{lm' - l'm}.$$

These equations, combined with
$$a\alpha + b\beta + c\gamma = 2\Delta,$$
give the values of α, β, γ, at the point of intersection.

10. *To find the equation of the straight line, passing through two given points.*

Let $f, g, h; f', g', h'$, be the co-ordinates of the two given points, and suppose the equation of the required straight line to be
$$L\alpha + M\beta + N\gamma = 0.$$

We must then have
$$Lf + Mg + Nh = 0,$$
$$Lf' + Mg' + Nh' = 0;$$
whence
$$\frac{L}{gh' - g'h} = \frac{M}{hf' - h'f} = \frac{N}{fg' - f'g},$$
giving, as the equation of the required line,
$$(gh' - g'h)\alpha + (hf' - h'f)\beta + (fg' - f'g)\gamma = 0.$$

11. *To find the general equation of a straight line, passing through the point of intersection of two given straight lines.*

If the equations of the straight lines be
$$l\alpha + m\beta + n\gamma = 0,$$
$$l'\alpha + m'\beta + n'\gamma = 0,$$
every straight line, passing through their point of intersection, may be represented by an equation of the form
$$l\alpha + m\beta + n\gamma = k\,(l'\alpha + m'\beta + n'\gamma),$$
where k is an arbitrary constant. For this equation is satisfied when the equations of the given straight lines are both satisfied, and, being of the first degree, it represents a straight line. It is therefore the equation of a straight line passing through their point of intersection.

12. *To find the condition that three points may lie in the same straight line.*

Let $\alpha_1, \beta_1, \gamma_1$; $\alpha_2, \beta_2, \gamma_2$; $\alpha_3, \beta_3, \gamma_3$, be the co-ordinates of the three given points, then, if these points lie in the same straight line, suppose the equation of that line to be
$$\lambda\alpha + \mu\beta + \nu\gamma = 0.$$
Then λ, μ, ν must satisfy the following equations:
$$\lambda\alpha_1 + \mu\beta_1 + \nu\gamma_1 = 0,$$
$$\lambda\alpha_2 + \mu\beta_2 + \nu\gamma_2 = 0,$$
$$\lambda\alpha_3 + \mu\beta_3 + \nu\gamma_3 = 0,$$
whence, eliminating λ, μ, ν, by cross multiplication,
$$\alpha_1\beta_2\gamma_3 - \alpha_1\beta_3\gamma_2 + \alpha_2\beta_3\gamma_1 - \alpha_2\beta_1\gamma_3 + \alpha_3\beta_1\gamma_2 - \alpha_3\beta_2\gamma_1 = 0,$$
the required condition.

13. *To find the condition that three straight lines may intersect in a point.*

Let the equations of the straight lines be
$$l_1\alpha + m_1\beta + n_1\gamma = 0,$$
$$l_2\alpha + m_2\beta + n_2\gamma = 0,$$
$$l_3\alpha + m_3\beta + n_3\gamma = 0.$$

If these three straight lines intersect in a point, the above three equations must be satisfied by the same values of α, β, γ. This gives, eliminating α, β, γ by cross-multiplication,
$$l_1 m_2 n_3 - l_1 m_3 n_2 + l_2 m_3 n_1 - l_2 m_1 n_3 + l_3 m_1 n_2 - l_3 m_2 n_1 = 0,$$
the required condition.

The identity of form between the conditions that three straight lines should intersect in a point, and that three points should lie in a straight line, is worthy of notice. Its full geometrical meaning will be seen hereafter.

We shall sometimes, in future investigations, speak of the straight line represented by the equation $l\alpha + m\beta + n\gamma = 0$, as the straight line (l, m, n). Adopting this phraseology, it will be seen that *the condition that the three points* (l_1, m_1, n_1) (l_2, m_2, n_2) (l_3, m_3, n_3) *should lie in the same straight line, is the same as the condition that the three straight lines* (l_1, m_1, n_1) (l_2, m_2, n_2) (l_3, m_3, n_3) *should intersect in a point.*

14. *To find the condition that two straight lines may be parallel to one another.*

Let the equations of the two straight lines be
$$l\alpha + m\beta + n\gamma = 0 \dots\dots\dots\dots\dots(1),$$
$$l'\alpha + m'\beta + n'\gamma = 0 \dots\dots\dots\dots\dots(2).$$

Let (f, g, h) (α, β, γ) be the co-ordinates of any two points in (1),

(f', g', h') $(\alpha', \beta', \gamma')$ be the co-ordinates of any two points in (2).

CONDITION OF PARALLELISM.

Then the condition of parallelism requires that

$$\frac{\alpha - f}{\alpha' - f'} = \frac{\beta - g}{\beta' - g'} = \frac{\gamma - h}{\gamma' - h'}.$$

Also, recurring to the investigation of Art. (8), fig. 7,

$$\frac{\alpha - f}{bn - cm} = \frac{\beta - g}{cl - an} = \frac{\gamma - h}{am - bl},$$

$$\frac{\alpha' - f'}{bn' - cm'} = \frac{\beta' - g'}{cl' - an'} = \frac{\gamma' - h'}{am' - bl'}.$$

Hence, the required condition of parallelism is

$$\frac{bn - cm}{bn' - cm'} = \frac{cl - an}{cl' - an'} = \frac{am - bl}{am' - bl'} \quad \ldots\ldots\ldots(3).$$

These two equations are equivalent to one only, since they may be written in the form

$$\frac{\dfrac{n}{c} - \dfrac{m}{b}}{\dfrac{n'}{c} - \dfrac{m'}{b}} = \frac{\dfrac{l}{a} - \dfrac{n}{c}}{\dfrac{l'}{a} - \dfrac{n'}{c}} = \frac{\dfrac{m}{b} - \dfrac{l}{a}}{\dfrac{m'}{b} - \dfrac{l'}{a}},$$

where it will be seen that the equality of any two members implies that the third is equal to either of them.

Multiplying the numerators and denominators of the several members of (3) by l', m', n' and adding, we obtain the condition under the form

$$(mn' - m'n)\,a + (nl' - n'l)\,b + (lm' - l'm)\,c = 0 \ldots\ldots(4).$$

This is the necessary condition of parallelism, and is generally the most convenient form which can be employed. It is equivalent to

$$(mn' - m'n)\sin A + (nl' - n'l)\sin B + (lm' - l'm)\sin C = 0,$$

a form which we shall occasionally use.

It will be observed that this condition is the same in form as that which results from the elimination of α, β, γ between the equations

$$l\alpha + m\beta + n\gamma = 0,$$
$$l'\alpha + m'\beta + n'\gamma = 0,$$
$$a\alpha + b\beta + c\gamma = 0.$$

The last of these is, as we know, an equation which cannot be satisfied by any values of α, β, γ, since, as we have already proved (Art. 2), $a\alpha + b\beta + c\gamma = 2\Delta$. Hence the equation (4) may be looked upon as an expression of the fact that the two equations

$$l\alpha + m\beta + n\gamma = 0,$$
$$l'\alpha + m'\beta + n'\gamma = 0,$$

cannot be simultaneously satisfied by any values of α, β, γ, or, in other words, that the two straight lines represented by them do not intersect, which is known to be a necessary condition of their parallelism, and also a sufficient condition, since the two straight lines are in the same plane.

Although, however, no values of α, β, γ exist which will satisfy the equation $a\alpha + b\beta + c\gamma = 0$, yet we can always satisfy the equation $l\alpha + m\beta + n\gamma = 0$, where the ratios $l:m:n$ approach as nearly as we please to the ratios $a:b:c$.

By referring to the investigation of Art. (7) it will be seen that, q, r, denoting the distances from A, of the points in which the straight line (l, m, n) cuts AC, AB respectively,

$$\frac{qra}{bc} = l, \quad \frac{qr}{c} - q = m, \quad \frac{qr}{b} - r = n;$$

whence
$$\frac{b}{a} - \frac{bc}{ar} = \frac{m}{l}, \quad \frac{c}{a} - \frac{bc}{aq} = \frac{n}{l},$$

or
$$\frac{c}{r} = 1 - \frac{am}{bl}, \quad \frac{b}{q} = 1 - \frac{an}{cl}.$$

It hence appears, that by making the ratios $l:m:n$ sufficiently nearly equal to the ratios $a:b:c$, the values of q and r may be made as great as we please, in other words, that the straight line (l, m, n) may be removed as far as we please from the triangle of reference. *The limiting position,*

CONDITION OF PARALLELISM.

therefore, *to which the straight line* (l, m, n) *continually approaches, and with which it ultimately coincides, when the ratios* l : m : n *continually approach to, and ultimately coincide with, the ratios* a : b : c, *is a straight line altogether at an infinite distance.*

This is often expressed by saying that the equation

$$a\alpha + b\beta + c\gamma = 0,$$

or the equivalent equation

$$\alpha \sin A + \beta \sin B + \gamma \sin C = 0,$$

represents *the straight line at infinity.*

This phraseology is very convenient, and free from objection, if the conventions on which it is adopted be clearly understood. It is, however, desirable that attention should be called to the fact, that the equation

$$a\alpha + b\beta + c\gamma = 0$$

is, in itself, *impossible*,—in fact, a contradiction in terms,— and can only be admitted as a limiting form to which possible equations may continually tend.

15. *To find the equation of a straight line, drawn through a given point, parallel to a given straight line.*

Let (l, m, n) be the given straight line, (f, g, h) the given point, then the equation of the required straight line will be

$$\frac{l\alpha + m\beta + n\gamma}{lf + mg + nh} = \frac{a\alpha + b\beta + c\gamma}{af + bg + ch}.$$

For this straight line passes through the point (f, g, h), and does not intersect the straight line (l, m, n); since, if it did, we should have $a\alpha + b\beta + c\gamma = 0$.

Since $af + bg + ch = 2\Delta$, this equation may also be written

$$l\alpha + m\beta + n\gamma = \frac{lf + mg + nh}{2\Delta}(a\alpha + b\beta + c\gamma).$$

F.   2

TRILINEAR CO-ORDINATES.

COR. The general equation of a straight line parallel to (l, m, n) is

$$l\alpha + m\beta + n\gamma = k(a\alpha + b\beta + c\gamma),$$

where k is an arbitrary constant.

16. *To find the inclinations of a straight line, drawn through one of the angular points of the triangle of reference, to the sides which intersect in that point.*

Let the equation of the straight line AP be

$$\mu\beta = \nu\gamma,$$

Fig. 8.

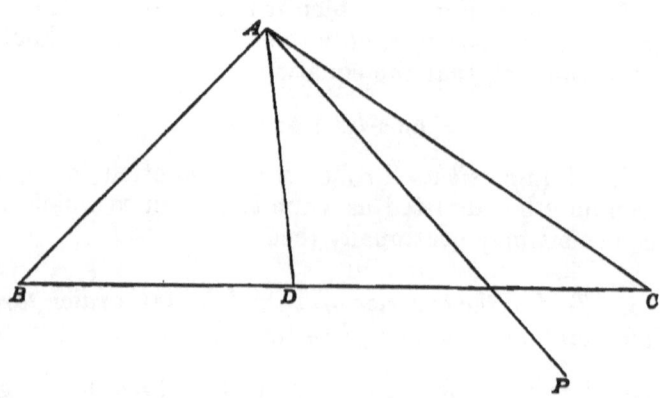

and let θ be its inclination to AD, the internal bisector of the angle A.

Then
$$\frac{\sin\left(\frac{A}{2} + \theta\right)}{\sin\left(\frac{A}{2} - \theta\right)} = \frac{\gamma}{\beta} = \frac{\mu}{\nu},$$

whence
$$\frac{\tan\theta}{\tan\frac{A}{2}} = \frac{\mu - \nu}{\mu + \nu};$$

CONDITION OF PERPENDICULARITY.

$$\therefore \tan\left(\theta + \frac{A}{2}\right) = \frac{\left(1 + \frac{\mu-\nu}{\mu+\nu}\right)\tan\frac{A}{2}}{1 - \frac{\mu-\nu}{\mu+\nu}\cdot\tan^2\frac{A}{2}}$$

$$= \frac{\mu \sin A}{\nu + \mu \cos A}.$$

Similarly, $\quad \tan\left(\theta - \dfrac{A}{2}\right) = -\dfrac{\nu \sin A}{\mu + \nu \cos A}.$

Hence, the inclinations of the given straight line to AB, AC, are determined.

17. *To find the condition that two given straight lines may be perpendicular to one another.*

Let (l, m, n), (l', m', n') be the two given straight lines. Through A draw two straight lines parallel to them. These will be represented by the equations

$$(ma - lb)\beta + (na - lc)\gamma = 0,$$
$$(m'a - l'b)\beta + (n'a - l'c)\gamma = 0.$$

And these straight lines must be at right angles to one another.

If θ, θ' be the respective inclinations of these straight lines to the internal bisector of the angle A, then, by the result of the last article,

$$\tan \theta = \frac{(lc - na) - (ma - lb)}{(lc - na) + (ma - lb)} \tan \frac{A}{2},$$

$$\tan \theta' = \frac{(l'c - n'a) - (m'a - l'b)}{(l'c - n'a) + (m'a - l'b)} \tan \frac{A}{2}.$$

And, if these be at right angles to one another,

$$1 + \tan \theta \tan \theta' = 0.$$

Hence

$$(lc - na)(l'c - n'a) + (ma - lb)(m'a - l'b)$$
$$+ \{(lc - na)(m'a - l'b) + (ma - lb)(l'c - n'a)\} \cos A = 0;$$

2—2

$$\therefore ll'(b^2 + c^2 - 2bc \cos A) + mm'a^2 + nn'a^2$$
$$- (mn' + m'n) a^2 \cos A - (nl' + n'l)(ac - ab \cos A)$$
$$- (lm' + l'm)(ab - ac \cos A) = 0,$$

which, since $b^2 + c^2 - 2bc \cos A = a^2$, $c - b \cos A = a \cos B$, $b - c \cos A = a \cos C$, reduces to

$$ll' + mm' + nn' - (mn' + m'n)\cos A - (nl' + n'l)\cos B - (lm' + l'm)\cos C = 0,$$

the required condition.

18. *To find the perpendicular distance from a given point to a given straight line.*

Let (f, g, h) be the given point, (l, m, n) the given straight line. Then, if q and r be the distance from A, of the points where this straight line meets AC, AB, respectively, we have shewn (Art. 7) that

$$\frac{1}{q} = \frac{1}{b} - \frac{na}{lbc},$$

$$\frac{1}{r} = \frac{1}{c} - \frac{ma}{lbc}.$$

Now, let a' denote the distance from (f, g, h) to (l, m, n). Then

$$(q^2 + r^2 - 2qr \cos A)^{\frac{1}{2}} a' + qg + rh = \frac{qr}{bc}(af + bg + ch),$$

or $\left(\dfrac{1}{q^2} + \dfrac{1}{r^2} - \dfrac{2\cos A}{qr}\right)^{\frac{1}{2}} a' = \dfrac{af + bg + ch}{bc} - \dfrac{g}{r} - \dfrac{h}{q}$

$$= \frac{a}{bc} f + \left(\frac{1}{c} - \frac{1}{r}\right) g + \left(1 - \frac{1}{q}\right) h$$

$$= \frac{a}{lbc}(lf + mg + nh).$$

And from the values of q and r

$$\frac{1}{q} - \frac{\cos A}{r} = \frac{1}{b} - \frac{\cos A}{c} - (n - m \cos A)\frac{a}{lbc}$$

$$= \frac{a(l \cos B + m \cos A - n)}{lbc}.$$

Similarly $\dfrac{1}{r} - \dfrac{\cos A}{q} = \dfrac{a(l\cos C + n\cos A - m)}{lbc};$

$\therefore \dfrac{1}{q^2} + \dfrac{1}{r^2} - \dfrac{2\cos A}{qr} = \left\{ \dfrac{a}{l^2 b^2 c^2} (lc - na)(l\cos B + m\cos A - n) \right.$

$\left. + (lb - ma)(l\cos C + n\cos A - m) \right\}$

$= \dfrac{a}{l^2 b^2 c^2} \{ l^2(c\cos B + b\cos C) + m^2 a + n^2 a - 2mn\, a\cos A$

$- nl(c + a\cos B - b\cos A) - lm(b - c\cos A + a\cos C)\},$

which, by reduction, is equal to

$\dfrac{a^2}{l^2 b^2 c^2} (l^2 + m^2 + n^2 - 2mn\cos A - 2nl\cos B - 2lm\cos C).$

Hence

$$a' = \pm \dfrac{lf + mg + nh}{(l^2 + m^2 + n^2 - 2mn\cos A - 2nl\cos B - 2lm\cos C)^{\frac{1}{2}}},$$

the required expression.

It will be observed, that the numerator of this expression vanishes if the point (f, g, h) lie upon the line (l, m, n), as manifestly ought to be the case.

It will also be remarked, that the more nearly the ratios $l : m : n$ approach to the ratios $a : b : c$, the less does the denominator of the above fraction become, and the greater, therefore, the distance from the point to the line; which is in accordance with the remark made in Art. (14).

EXAMPLES.

1. Find the equation of the straight line joining the middle points of two sides of the triangle of reference; and thence prove that it is parallel to the third side.

2. Find the equations of the straight lines, drawn through the several angular points of the triangle of reference, respectively at right angles to

$$\frac{\beta}{b} + \frac{\gamma}{c} = 0, \quad \frac{\gamma}{c} + \frac{a}{a} = 0, \quad \frac{a}{a} + \frac{\beta}{b} = 0;$$

and thence prove that they intersect in a point.

3. If θ be the angle between the two straight lines (l, m, n), (λ, μ, ν), prove that

$$\cot\theta = \frac{l\lambda + m\mu + n\nu - (m\nu + n\mu)\cos A - (n\lambda + l\nu)\cos B - (l\mu + m\lambda)\cos C}{(m\nu - n\mu)\sin A + (n\lambda - l\nu)\sin B + (l\mu - m\lambda)\sin C}.$$

4. On the sides of the triangle ABC, as bases, are constructed three triangles $A'BC, AB'C, ABC'$, similar to each other, and so placed that the angle $BA'C = B'AC = BAC'$, $CB'A = C'BA = CBA'$, $AC'B = A'CB = ACB'$. Prove that the straight lines AA', BB', CC' intersect in one point.

5. Prove that the straight line, joining the centre of the circle inscribed in the triangle ABC, with the middle point of the side BC, is parallel to the straight line joining A with the point of contact of the circle touching BC externally and AB, AC produced.

6. On the sides BC, CA, AB of the triangle ABC, respectively, pairs of points are taken, B_1, C_1; C_2, A_2; A_3, B_3; such that the points of intersection of BC with B_3C_2, of CA with C_1A_3, and of AB with A_2B_1 lie in a straight line; BC_2, CB_3 intersect in L; CA_3, AC_1 in M; AB_1, BA_2 in N. Prove that AL, BM, CN intersect in one point.

7. From the vertices of a triangle ABC, three straight lines AP, BQ, CR are drawn to pass through one point, and three straight lines AP', BQ', CR' to pass through another point, the points P, P' lying on BC, Q, Q' on CA, R, R' on AB; BQ, CR meet AP' in D_1, D_2; CR, AP meet BQ' in E_1, E_2; AP, BQ meet CR' in F_1, F_2; CD_1, BD_2 intersect in L; AE_1, CE_2 in M; BF_1, AF_2 in N. Prove that AL, BM, CN intersect in a point.

Anharmonic Ratio.

19. We shall introduce, in this place, a short account of harmonic and anharmonic section, as a familiarity with this conception is useful in the higher geometrical investigations.

Def. 1. If OP, OQ, OR, OS be four straight lines intersecting in a point, the ratio

$$\frac{\sin POQ \cdot \sin ROS}{\sin POS \cdot \sin QOR}$$

is called the anharmonic ratio of the pencil OP, OQ, OR, OS, and is expressed by the notation $\{O \cdot PQRS\}$*.

Def. 2. If P, Q, R, S be four points in a straight line, the ratio $\dfrac{PQ \cdot RS}{PS \cdot QR}$ is called the anharmonic ratio of the range P, Q, R, S, and may be expressed thus $[PQRS]$.

In using these definitions, attention must be paid to the order in which the lines or points follow one another. Thus, the anharmonic ratio of the pencil OP, OR, OQ, OS, is different from that of the pencil OP, OQ, OR, OS, the former being equal to $\dfrac{\sin POR \cdot \sin QOS}{\sin POS \cdot \sin QOR}$, the latter to $\dfrac{\sin POQ \cdot \sin ROS}{\sin POS \cdot \sin QOR}$.

Def. 3. If any number of straight lines, intersecting in a point, be cut by another straight line, the straight line which cuts the others is called a *transversal*.

20. Prop. *If four given straight lines, intersecting in a point O, be cut by a transversal in the points* P, Q, R, S, *the anharmonic ratio of the pencil* OP, OQ, OR, OS, *will be equal to that of the range* P, Q, R, S.

---

* This notation is due, I believe, to Dr Salmon. See his *Conic Sections* p. 297 (sixth edition).

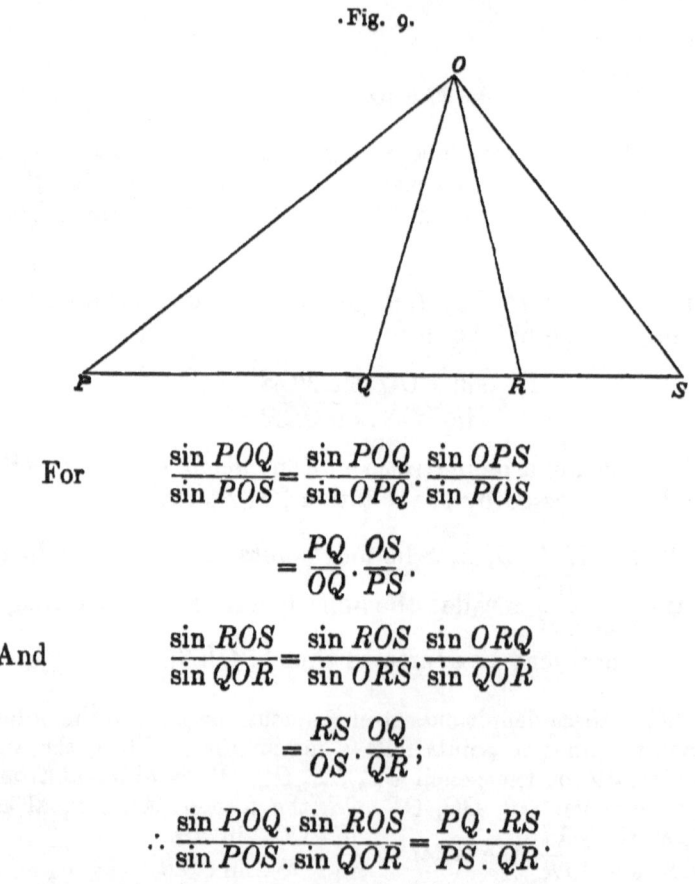

Fig. 9.

For
$$\frac{\sin POQ}{\sin POS} = \frac{\sin POQ}{\sin OPQ} \cdot \frac{\sin OPS}{\sin POS}$$

$$= \frac{PQ}{OQ} \cdot \frac{OS}{PS}.$$

And
$$\frac{\sin ROS}{\sin QOR} = \frac{\sin ROS}{\sin ORS} \cdot \frac{\sin ORQ}{\sin QOR}$$

$$= \frac{RS}{OS} \cdot \frac{OQ}{QR};$$

$$\therefore \frac{\sin POQ \cdot \sin ROS}{\sin POS \cdot \sin QOR} = \frac{PQ \cdot RS}{PS \cdot QR}.$$

Thus the proposition is proved.

COR. 1. It appears, from the above proposition, that if a pencil be cut by two distinct transversals in P, Q, R, S and P', Q', R', S' respectively, the anharmonic ratio of the range P, Q, R, S will be equal to that of the range P', Q', R', S', since each is equal to that of the pencil OP, OQ, OR, OS.

COR. 2. It appears also that, if four points P, Q, R, S, lying in a straight line, be joined with each of two other points O, O', the anharmonic ratios of the pencils OP, OQ,

OR, OS; $O'P$, $O'Q$, $O'R$, $O'S$, will be equal to one another, since each is equal to that of the range P, Q, R, S.

21. DEF. A pencil, of which the anharmonic ratio is unity, is called an harmonic pencil.

A range, of which the anharmonic ratio is unity, is called an harmonic range, and the straight line, on which the range lies, is said to be divided harmonically.

From what has been said above, it will be seen that, if an harmonic pencil be cut by a transversal, the four points of section will form an harmonic range. And if four points, forming an harmonic range, be joined with a fifth point, the four joining lines will form an harmonic pencil.

The line OS is said to be a fourth harmonic to the pencil OP, OQ, OR; and the point S to be a fourth harmonic to the range P, Q, R.

The term *harmonic* is employed on account of the circumstance, that if the points P, Q, R, S form what is above defined as an harmonic range, PR will be an harmonic mean between PQ and PS.

For $$PQ \cdot RS = PS \cdot QR\,;$$
$$\therefore PQ\,(PS - PR) = PS\,(PR - PQ)\,;$$
$$\therefore PQ : PS :: PR - PQ : PS - PR,$$

whence PQ, PR, PS are in harmonical progression.

From the above proportion it appears that if $PQ = QR$, $PS = \infty$. Hence, if PR be bisected in Q, the fourth harmonic to the range P, Q, R is infinitely distant. Or, as it may otherwise be stated, if PR be bisected in Q, and P, Q, R be joined with any point O, not in the line PR, the fourth harmonic to the pencil OP, OQ, OR, will be parallel to the transversal PQR.

22. PROP. *The external and internal bisectors of any angle form, with the lines containing the angle, an harmonic pencil.*

TRILINEAR CO-ORDINATES.

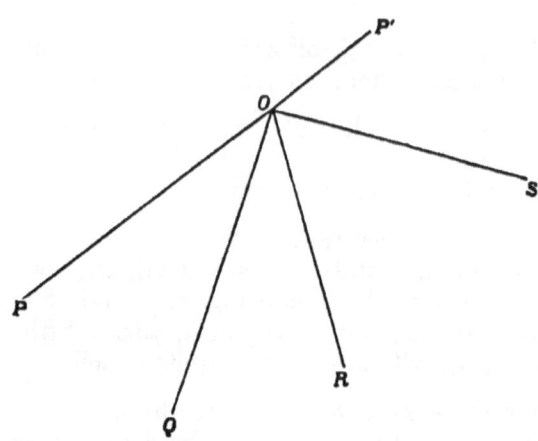

Fig. 10.

Let the angle POR be bisected internally by OQ, let PO be produced to any point P', and let the angle $P'OR$ be bisected by OS, then

$$\sin POQ = \sin QOR,$$

$$\sin POS = \sin P'OS$$

$$= \sin ROS;$$

$$\therefore \frac{\sin POQ . \sin ROS}{\sin POS . \sin QOR} = 1.$$

Hence the truth of the proposition.

23. PROP. *If* ABC *be the triangle of reference, and* AD, AE *straight lines respectively represented by the equations*

$$\beta - k\gamma = 0, \quad \beta + k'\gamma = 0,$$

then $\dfrac{k'}{k}$ *will be the anharmonic ratio of the pencil* AB, CA, AD, AE.

HARMONIC PENCILS.

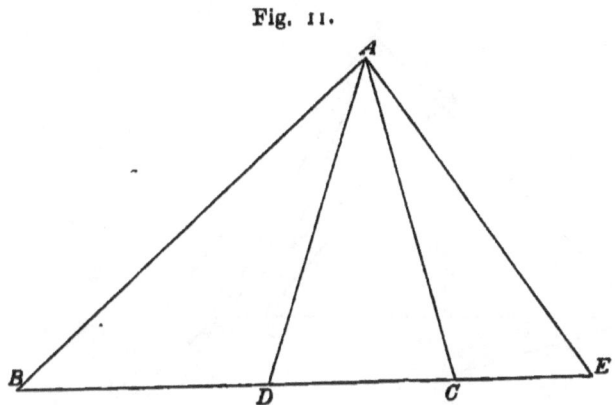

Fig. 11.

Let BC cut AD, AE respectively in D, E, then since D is a point in the line $\beta - k\gamma = 0$,

$$\frac{BD}{CD} = \frac{\triangle ABD}{\triangle ACD} = \frac{c\gamma}{b\beta} = \frac{c}{bk},$$

and since E is a point in the line $\beta + k'\gamma = 0$,

$$\frac{BE}{CE} = \frac{\triangle ABE}{\triangle ACE} = \frac{c\gamma}{-b\beta} = \frac{c}{bk'};$$

$$\therefore \frac{BD \cdot CE}{BE \cdot CD} = \frac{k'}{k},$$

or $\dfrac{k'}{k}$ is the anharmonic ratio of the range B, D, C, E; that is, of the pencil AB, AD, AC, AE.

COR. It hence follows that the straight lines respectively represented by the equations $\beta = 0$, $\beta - k\gamma = 0$, $\gamma = 0$, $\beta + k\gamma = 0$, form an harmonic pencil.

24. Hence we deduce a geometrical construction for the determination of the fourth harmonic to three given intersecting straight lines.

Let AB, AD, AC be three given intersecting straight lines, and let it be required to find a straight line AE, such that AB, AD, AC, AE shall form an harmonic pencil.

Fig. 12.

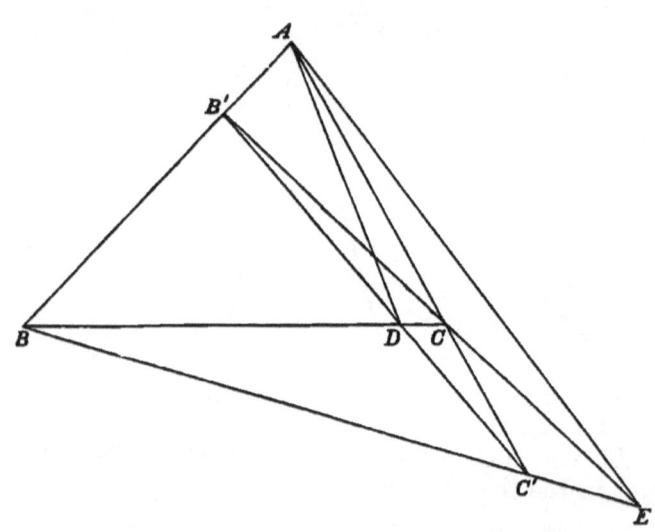

Through D, any point of the second of the three given straight lines, draw two transversals BDC, $B'DC'$, cutting AB in B, B', AC in C, C' respectively. Join $B'C$, BC', and produce them to meet in E. Join AE, then AE shall be the fourth harmonic required.

For, let ABC be the triangle of reference, and let the equation of AD be $\beta - k\gamma = 0$. Let the equation of $B'C'$ be $\lambda\alpha + \beta - k\gamma = 0$.

Then that of BC' is $\lambda\alpha - k\gamma = 0$,

.... $B'C$ ... $\lambda\alpha + \beta = 0$,

∴ .... AE ... $\beta + k\gamma = 0$,

whence AE is the fourth harmonic required.

25. PROP. *If* ABC *be a given triangle,* P *any given point; and* AD, *the fourth harmonic to* AB, AP, AC, *intersect* BC *in* D; BE, *the fourth harmonic to* BC, BP, BA,

HARMONIC PENCILS.

intersect CA *in* E; CF, *the fourth harmonic to* CA, CP, CB, *intersect* AB *in* F; *then* D, E, F *lie in the same straight line*.

Let f, g, h be the co-ordinates of P. Then the equation of AP is

$$\frac{\beta}{g} - \frac{\gamma}{h} = 0,$$

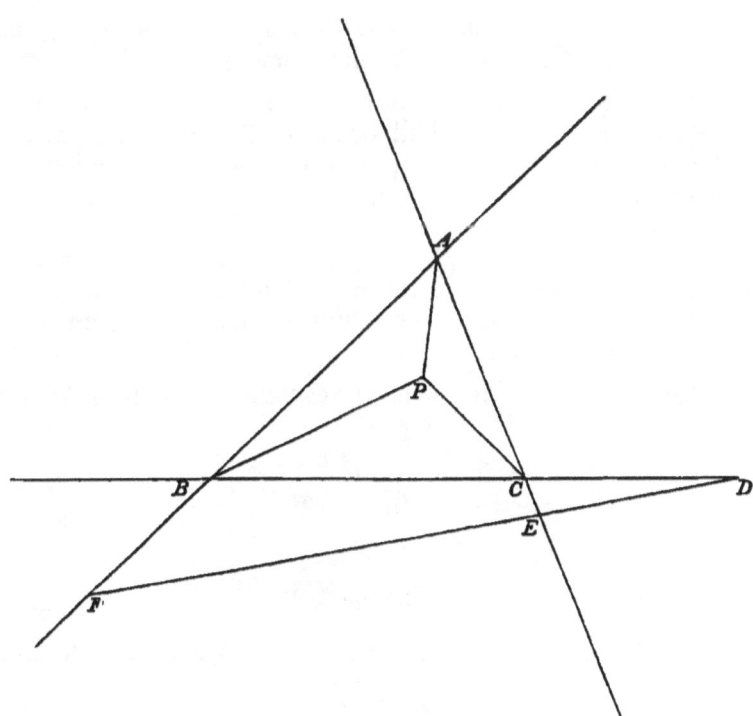

Fig. 13.

whence that of AD is $\quad \dfrac{\beta}{g} + \dfrac{\gamma}{h} = 0.$

Similarly, that of BE is $\dfrac{\gamma}{h} + \dfrac{\alpha}{f} = 0,$

..... CF ... $\dfrac{\alpha}{f} + \dfrac{\beta}{g} = 0.$

From the form of these equations it will be seen that the straight line

$$\frac{\alpha}{f}+\frac{\beta}{g}+\frac{\gamma}{h}=0$$

passes through D, E, F. Hence these three points are in a straight line.

COR. The converse proposition to that above enunciated may be demonstrated by similar reasoning.

The point P, and the line DEF, may be called *harmonics* of one another with respect to the triangle ABC.

By combining the proposition last proved with that proved in Art. (22), we shall obtain a demonstration of the statements made in Art. 6; that the points in which the external bisectors of each angle of a triangle respectively intersect the sides opposite to them, lie in the same straight line; and that the points in which the external bisector of any one angle and the internal bisectors of the other two angles, intersect the sides respectively opposite to them, lie in the same straight line.

These straight lines will be respectively represented by the equations,

$$\alpha+\beta+\gamma=0, \quad \beta+\gamma-\alpha=0,$$
$$\gamma+\alpha-\beta=0, \quad \alpha+\beta-\gamma=0.$$

ON INVOLUTION.

26. DEFS. Let O be a point in a given straight line, and let

$$P, P', Q, Q', R, R'......$$

be a series of points on that line so taken that

$$OP.OP'=OQ.OQ'=OR.OR'=......$$

$$= \text{a constant, } k^2 \text{ suppose.}$$

Then these points are said to form a system in *involution*.

If K be a point such that $OK^2=k^2$, K is called a *focus* of the system.

If k^2 be positive, there will evidently be two such foci, one on each side of O, if negative (and k therefore imaginary) there will be no real foci.

The point O is called the *centre* of the system.

Two points, such as P, P', are said to be *conjugate* to one another.

It is evident that each focus is conjugate to itself, and that the conjugate of the centre is at an infinite distance, and that a point and its conjugate will be on the same, or different sides of the centre, according as the foci are real or imaginary.

The system will be determined when two foci, or a centre and focus, are given. It will also be determined if two pair of conjugate points be given; as may be seen as follows.

Let p, p', q, q' be the respective distances of the four points from any arbitrary point on the line, x the distance of the centre from the same point.

Then, by definition,
$$(p-x)(p'-x) = (q-x)(q'-x);$$
$$\therefore x = \frac{pp' - qq'}{p + p' - q - q'},$$
which determines the centre.

27. Prop. *The anharmonic ratio of four points is equal to that of their four conjugates.*

For, if $OP = p$, $OQ = q$, $OR = r$, $OS = s$,
$$\text{then } [PQRS] = \frac{(q-p)(s-r)}{(s-p)(r-q)},$$
$$\text{and } [P'Q'R'S'] = \frac{\left(\dfrac{k^2}{q} - \dfrac{k^2}{p}\right)\left(\dfrac{k^2}{s} - \dfrac{k^2}{r}\right)}{\left(\dfrac{k^2}{s} - \dfrac{k^2}{p}\right)\left(\dfrac{k^2}{r} - \dfrac{k^2}{q}\right)}$$
$$= \frac{(p-q)(r-s)}{(p-s)(q-r)}$$
$$= [PQRS],$$
which proves the proposition.

COR. It is evident that $[PQRP'] = [P'Q'R'P]$.

28. PROP. *Any two conjugate points form, with the two foci, an harmonic range.*

Let K_1, K_2 be the foci, then

$$K_1P = p - k, \quad K_2P = p + k,$$

$$K_1P' = \frac{k^2}{p} - k, \quad K_2P' = \frac{k^2}{p} + k,$$

then $K_1P \cdot K_2P' = (p - k)\left(\frac{k^2}{p} + k\right) = \frac{k}{p}(p^2 - k^2),$

and $K_1P' \cdot K_2P = \left(k - \frac{k^2}{p}\right)(k + p) = \frac{k}{p}(p^2 - k^2)$;

$$\therefore K_1P \cdot K_2P' = K_1P' \cdot K_2P,$$

or the four points in question form a harmonic range.

Conversely, if there be a system of pairs of points in a straight line, such that each pair forms, with two given points, an harmonic range, the aggregate of the pairs of points will form a system in involution, of which the two given points are the foci.

29. A system of straight lines, intersecting in a point, may be treated in the same manner as a system of points lying in a straight line, the sine of the angle between any two lines taking the place of the mutual distance of two points. From the proposition, proved in Art. 20, it will follow that, if a system of straight lines in involution be cut by a transversal, the points of section will also be in involution.

CHAPTER II.

SPECIAL FORMS OF THE EQUATION OF THE SECOND DEGREE.

1. WE now proceed to the discussion of the curve represented by the equation of the second degree. We shall first prove that every curve, represented by such an equation, is what is commonly called a conic section; and then, before proceeding further with the consideration of the general equation, shall investigate the nature of the curve corresponding to certain special forms of the equation.

PROP. *Every curve represented by an equation of the second degree is cut by a straight line in two points, real, coincident, or imaginary.*

The general equation of the second degree is represented by

$$u\alpha^2 + v\beta^2 + w\gamma^2 + 2u'\beta\gamma + 2v'\gamma\alpha + 2w'\alpha\beta = 0.$$

To find where the curve, of which this is the equation, is cut by the straight line

$$l\alpha + m\beta + n\gamma = 0,$$

we may eliminate α between the two equations. This will give us a quadratic for the determination of $\frac{\beta}{\gamma}$, to each of the two values of this ratio, real, equal, or imaginary, one value of α will correspond; whence it appears that the straight line and the curve cut one another in two real, coincident, or imaginary points.

Hence, the curve is of the same nature as that represented by the equation of the second degree in Cartesian co-ordinates, and is, therefore, a conic section.

F.

2. We shall now inquire what are the relations of the conic section to the triangle of reference, when certain relations exist among the coefficients of the equation.

First, suppose u, v, w, all $= 0$.

The equation then assumes the form
$$u'\beta\gamma + v'\gamma\alpha + w'\alpha\beta = 0,$$
which we shall write
$$\lambda\beta\gamma + \mu\gamma\alpha + \nu\alpha\beta = 0.$$

Now, if in this equation we put $\alpha = 0$, it reduces itself to
$$\lambda\beta\gamma = 0,$$
which requires either that $\beta = 0$, or that $\gamma = 0$.

It hence appears that the curve passes through two of the angular points (B, C) of the triangle of reference. It may similarly be shewn to pass through the third. Hence the equation
$$\lambda\beta\gamma + \mu\gamma\alpha + \nu\alpha\beta = 0,$$
or, as it may also be written,
$$\frac{\lambda}{\alpha} + \frac{\mu}{\beta} + \frac{\nu}{\gamma} = 0,$$
represents *a conic, described about the triangle of reference.*

3. Let us now inquire how the line
$$\frac{\beta}{\mu} + \frac{\gamma}{\nu} = 0$$
is related to this conic.

If in the equation of the conic we put $\frac{\beta}{\mu} + \frac{\gamma}{\nu} = 0$, or, which is the same thing, $\mu\gamma + \nu\beta = 0$, it reduces to $\lambda\beta\gamma = 0$. Hence the line $\frac{\beta}{\mu} + \frac{\gamma}{\nu} = 0$ meets the conic in the points in which it meets the lines $\beta = 0$, $\gamma = 0$; but these two points coincide, since the line in question evidently passes through the point of intersection of $\beta = 0$ and $\gamma = 0$. Hence the straight line and the conic meet one another in coincident points, that is, they touch one another at the point A.

CENTRE OF THE CONIC.   35

Similarly, the equations of the tangents at B and C are

$$\frac{\gamma}{\nu} + \frac{\alpha}{\lambda} = 0,$$

$$\frac{\alpha}{\lambda} + \frac{\beta}{\mu} = 0.$$

4. *To determine the position of the centre of the conic.*

Through the angular points A, B, C of the triangle of reference draw the tangents EAF, FBD, DCE. Bisect

Fig. 14.

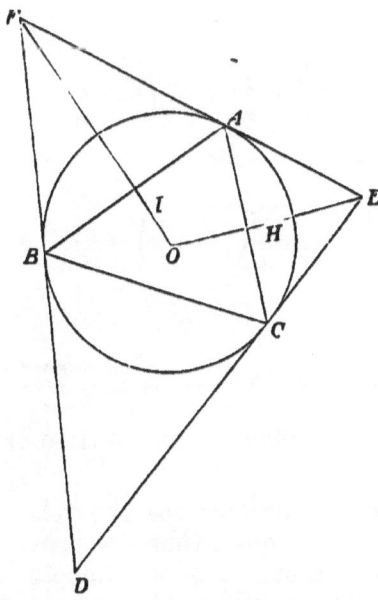

AC, AB respectively in H, I, join EH, FI, and produce them to intersect in O. Then, since every straight line drawn through the intersection of two tangents so as to bisect their chord of contact passes also through the centre, O will be the centre of the conic.

Now, at the point E, we have

$$\frac{\alpha}{\lambda} = -\frac{\beta}{\mu} = \frac{\gamma}{\nu},$$

and at the point H,

$$\beta = 0, \quad c\gamma = a\alpha.$$

Hence the equation of EH is

$$\lambda a \left(\frac{\alpha}{\lambda} + \frac{\beta}{\mu}\right) = \nu c \left(\frac{\beta}{\mu} + \frac{\gamma}{\nu}\right),$$

or $\quad \dfrac{1}{\nu c}\left(\dfrac{\alpha}{\lambda} + \dfrac{\beta}{\mu}\right) = \dfrac{1}{\lambda a}\left(\dfrac{\beta}{\mu} + \dfrac{\gamma}{\nu}\right).$

Similarly that of FI is

$$\frac{1}{\lambda a}\left(\frac{\beta}{\mu} + \frac{\gamma}{\nu}\right) = \frac{1}{\mu b}\left(\frac{\gamma}{\nu} + \frac{\alpha}{\lambda}\right).$$

Hence, at the point O,

$$\frac{1}{\lambda a}\left(\frac{\beta}{\mu} + \frac{\gamma}{\nu}\right) = \frac{1}{\mu b}\left(\frac{\gamma}{\nu} + \frac{\alpha}{\lambda}\right) = \frac{1}{\nu c}\left(\frac{\alpha}{\lambda} + \frac{\beta}{\mu}\right),$$

or $\quad \dfrac{\frac{\alpha}{\lambda}}{-\lambda a + \mu b + \nu c} = \dfrac{\frac{\beta}{\mu}}{\lambda a - \mu b + \nu c} = \dfrac{\frac{\gamma}{\nu}}{\lambda a + \mu b - \nu c}.$

These equations determine the position of the centre.

Cor. We may hence deduce the relation which must hold between λ, μ, ν, in order that the conic may be a parabola. For, since the centre of a parabola is at an infinite distance, its co-ordinates will satisfy the equation

$$a\alpha + b\beta + c\gamma = 0.$$

We hence obtain the following equation:

$$\lambda^2 a^2 + \mu^2 b^2 + \nu^2 c^2 - 2\mu\nu bc - 2\nu\lambda ca - 2\lambda\mu ab = 0,$$

which is equivalent to

$$\pm (\lambda a)^{\frac{1}{2}} \pm (\mu b)^{\frac{1}{2}} \pm (\nu c)^{\frac{1}{2}} = 0,$$

as the necessary and sufficient condition that the conic should be a parabola.

5. *To determine the condition that a given straight line may touch the conic.*

If the conic be touched by the straight line (l, m, n), the two values of the ratio $\beta : \gamma$, obtained by eliminating α between the equations

$$\lambda\beta\gamma + \mu\gamma\alpha + \nu\alpha\beta = 0,$$

$$l\alpha + m\beta + n\gamma = 0,$$

must be coincident. The equation which determines these is

$$-\lambda l \beta\gamma + (\mu\gamma + \nu\beta)(m\beta + n\gamma) = 0,$$

and the condition that the two values of $\beta : \gamma$ be equal, is

$$4\mu n \cdot \nu m - (\mu m + \nu n - \lambda l)^2 = 0,$$

or $\quad \lambda^2 l^2 + \mu^2 m^2 + \nu^2 n^2 - 2\mu\nu \cdot mn - 2\nu\lambda \cdot nl - 2\lambda\mu \cdot lm = 0,$

which is equivalent to

$$\pm (\lambda l)^{\frac{1}{2}} \pm (\mu m)^{\frac{1}{2}} \pm (\nu n)^{\frac{1}{2}} = 0.$$

If this be compared with the condition investigated in Art. (4) that the conic may be a parabola, it will be observed that the parabola satisfies the *analytical condition* of touching the straight line $a\alpha + b\beta + c\gamma = 0$. This is generally expressed by saying that *every parabola touches the line at infinity.*

6. *To investigate the equation of the circle, circumscribing the triangle of reference.*

This may be deduced from the consideration that the co-ordinates of the centre of the circumscribing circle are respectively proportional to $\cos A$, $\cos B$, $\cos C$ (see p. 4). Or it may be independently investigated as follows. Draw EAF, FAD, DAE (fig. 2), tangents to the circle, then the angle EAC is equal to ABC, and FAB to ACB (Euc. III. 32). Hence the equation of the tangent EAF must be

$$\frac{\beta}{\sin B} + \frac{\gamma}{\sin C} = 0,$$

Fig. 15.

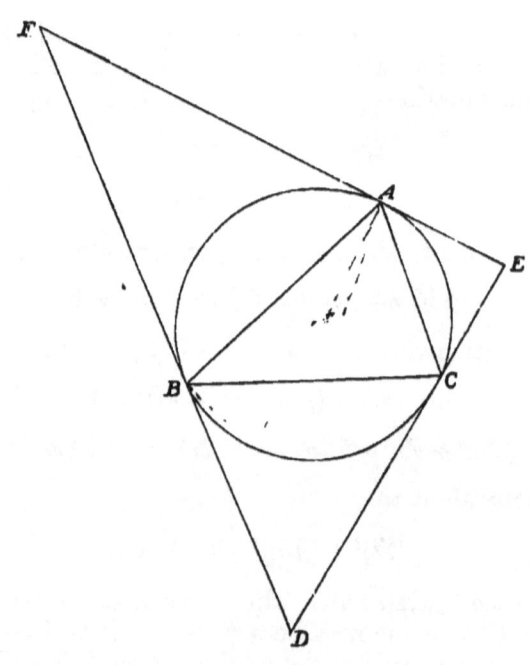

or $$\frac{\beta}{b} + \frac{\gamma}{c} = 0.$$

Similarly the equations of the other tangents FBD, DCE are

$$\frac{\gamma}{c} + \frac{\alpha}{a} = 0,$$

$$\frac{\alpha}{a} + \frac{\beta}{b} = 0;$$

and, comparing these with the forms of equations of the tangents given in Art. (3), we see that the equation of the circumscribing circle is

$$\frac{a}{\alpha} + \frac{b}{\beta} + \frac{c}{\gamma} = 0,$$

or, as it may also be written,

$$\frac{\sin A}{\alpha} + \frac{\sin B}{\beta} + \frac{\sin C}{\gamma} = 0.$$

7. Having thus discussed the equation of the conic, circumscribing the triangle of reference, we may proceed to investigate that of the conic which touches its three sides. The condition that the conic

$$u\alpha^2 + v\beta^2 + w\gamma^2 + 2u'\beta\gamma + 2v'\gamma\alpha + 2w'\alpha\beta = 0,$$

may touch the line $\alpha = 0$ is, that the left-hand member of the equation obtained by writing $\alpha = 0$ in the above may be a perfect square. This requires that

$$u'^2 = vw,$$

or

$$u' = \pm (vw)^{\frac{1}{2}}.$$

Similarly,

$$v' = \pm (wu)^{\frac{1}{2}},$$

$$w' = \pm (uv)^{\frac{1}{2}},$$

are necessary conditions that the conic should touch the lines $\beta = 0$, $\gamma = 0$.

We must observe, however, that if the conic touch all three of the sides of the triangle of reference, the three double signs in the above equations must be taken *all negatively*, or *two positively and one negatively*. For, if they be taken otherwise, the left-hand member of the equation of the conic will become a perfect square, as may be ascertained by substitution, and the conic will degenerate into a straight line, or rather into two coincident straight lines.

Taking then the double signs all negatively, and writing for convenience, L^2, M^2, N^2, instead of u, v, w, the equation of the conic which touches the three sides of the triangle of reference becomes

$$L^2\alpha^2 + M^2\beta^2 + N^2\gamma^2 - 2MN\beta\gamma - 2NL\gamma\alpha - 2LM\alpha\beta = 0,$$

which is equivalent to

$$\pm(L\alpha)^{\frac{1}{2}} \pm (M\beta)^{\frac{1}{2}} \pm (N\gamma)^{\frac{1}{2}} = 0.$$

It may be remarked, that the condition that the point (l, m, n) should lie in the above conic, is the same as the condition that the straight line (l, m, n) should touch the circumscribing conic

$$L\beta\gamma + M\gamma\alpha + N\alpha\beta = 0.$$

See Art. 5. This we shall return to hereafter.

8. *To find the centre of the conic.*

Let D, E, F be the points of contact of the sides BC, CA, AB respectively. Join EF, FD, DE, bisect FD, DE in H, I, join BH, CI, and produce them to meet in O. Then O will be the centre of the conic (see p. 32). We have then to find the equations of BH, CI, which, by their intersection, determine O.

Fig. 16.

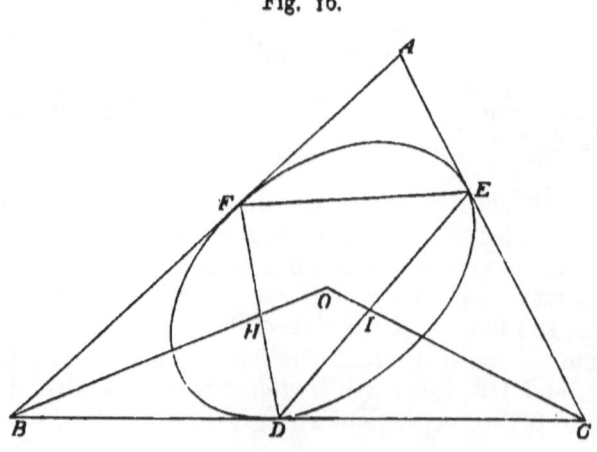

CO-ORDINATES OF THE CENTRE.

Let f_1, g_1, h_1 be the co-ordinates of D. Then $f_1 = 0$; and g_1, h_1 will be the values of β, γ, which satisfy the equations

$$M^2\beta^2 + N^2\gamma^2 - 2MN\beta\gamma = 0,$$

or

$$M\beta - N\gamma = 0,$$

and

$$b\beta + c\gamma = 2\Delta.$$

Hence,

$$g_1 = \frac{N}{Nb + Mc} 2\Delta$$

$$= \frac{\frac{1}{M}}{\frac{b}{M} + \frac{c}{N}} 2\Delta.$$

In like manner it may be proved that, if f_2, g_2, h_2 be the co-ordinates of E,

$$f_2 = \frac{\frac{1}{L}}{\frac{a}{L} + \frac{c}{N}} 2\Delta, \quad g_2 = 0.$$

Now, for I, and therefore for every point in the line CI,

$$\frac{\alpha}{\frac{1}{2}(f_1 + f_2)} = \frac{\beta}{\frac{1}{2}(g_1 + g_2)}.$$

Therefore the equation of CI is

$$L\alpha\left(\frac{a}{L} + \frac{c}{N}\right) = M\beta\left(\frac{b}{M} + \frac{c}{N}\right),$$

or

$$\frac{\alpha}{Nb + Mc} = \frac{\beta}{Lc + Na}.$$

Similarly that of BH is

$$\frac{\gamma}{Ma + Lb} = \frac{\alpha}{Nb + Mc}.$$

Therefore at the point O we have
$$\frac{\alpha}{Nb + Mc} = \frac{\beta}{Lc + Na} = \frac{\gamma}{Ma + Lb}.$$
These equations with
$$a\alpha + b\beta + c\gamma = 2\Delta$$
determine the co-ordinates of the centre.

Cor. Hence may be obtained the condition that the conic may be a parabola. For the centre of a parabola is infinitely distant, its co-ordinates must therefore satisfy the algebraical relation
$$a\alpha + b\beta + c\gamma = 0,$$
whence we get
$$Lbc + Mca + Nab = 0,$$
or
$$\frac{L}{a} + \frac{M}{b} + \frac{N}{c} = 0,$$
as the required condition.

This will be observed by reference to Art. 9, to be identical with the condition that the conic should touch the straight line, $a\alpha + b\beta + c\gamma = 0$, and thus we are again led to the conclusion noticed in Art. 7, that every parabola touches the line at infinity.

9. *To find the condition that the conic should touch a given straight line.*

If the straight line (l, m, n) be a tangent to the conic, the values of the ratio $\beta : \gamma$, obtained by eliminating α between the equation of the conic and the equation
$$l\alpha + m\beta + n\gamma = 0,$$
must be equal to one another. For this purpose, it is most convenient to take the equation of the conic in the form
$$\pm (L\alpha)^{\frac{1}{2}} \pm (M\beta)^{\frac{1}{2}} \pm (N\gamma)^{\frac{1}{2}} = 0.$$

Eliminating α, we then get

$$L(m\beta + n\gamma) + l\{(M\beta)^{\frac{1}{2}} \pm (N\gamma)^{\frac{1}{2}}\}^2 = 0,$$

or $(Lm + Ml)\beta + (Ln + Nl)\gamma \pm 2l(MN\beta\gamma)^{\frac{1}{2}} = 0,$

and, if the roots of this, considered as a quadratic in $\left(\dfrac{\beta}{\gamma}\right)^{\frac{1}{2}}$ be equal, we have

$$(Lm + Ml)(Ln + Nl) - l^2 MN = 0,$$

or $Lmn + Mnl + Nlm = 0,$

which may also be written

$$\frac{L}{l} + \frac{M}{m} + \frac{N}{n} = 0.$$

It hence appears that the condition, that the line (l, m, n) should touch the conic

$$(L\alpha)^{\frac{1}{2}} \pm (M\beta)^{\frac{1}{2}} + (N\gamma)^{\frac{1}{2}} = 0,$$

is identical with the condition that the point (l, m, n) should lie in the conic

$$\frac{L}{\alpha} + \frac{M}{\beta} + \frac{N}{\gamma} = 0;$$

a result analogous to that obtained in Art. 13, Chap. I.

10. *To find the equations of the four circles which touch the three sides of the triangle of reference.*

These may be obtained most readily by the employment of the equations for the determination of the centre, obtained in Art. 8. Thus, let it be required to find the ratios of L, M, N in order that the conic may become the *inscribed* circle. At the centre of this circle we have, as we know, $\alpha = \beta = \gamma$.

This gives, by the result of Art. 8,
$$Nb + Mc = Lc + Na = Ma + Lb.$$

To solve these equations, put each member equal to r, we then get
$$\frac{M}{b} + \frac{N}{c} = \frac{r}{bc},$$
$$\frac{N}{c} + \frac{L}{a} = \frac{r}{ca},$$
$$\frac{L}{a} + \frac{M}{b} = \frac{r}{ab}.$$

Adding together the last two of these equations, and subtracting the first, we get
$$L = \frac{r}{2} \frac{(b+c-a)}{bc}.$$

Similar expressions being obtained for M and N, we see that
$$L : M : N :: \frac{b+c-a}{bc} : \frac{c+a-b}{ca} : \frac{a+b-c}{ab}$$
$$:: \cos^2 \frac{A}{2} : \cos^2 \frac{B}{2} : \cos^2 \frac{C}{2}.$$

Hence the inscribed circle is represented by the equation
$$\cos \frac{A}{2} \cdot \alpha^{\frac{1}{2}} + \cos \frac{B}{2} \cdot \beta^{\frac{1}{2}} + \cos \frac{C}{2} \cdot \gamma^{\frac{1}{2}} = 0.$$

It may similarly be proved that the escribed circles, of which the centres are respectively given by
$$-\alpha = \beta = \gamma, \quad \alpha = -\beta = \gamma, \quad \alpha = \beta = -\gamma,$$
will be represented by the equations
$$\cos \frac{A}{2} (-\alpha)^{\frac{1}{2}} + \sin \frac{B}{2} \cdot \beta^{\frac{1}{2}} + \sin \frac{C}{2} \cdot \gamma^{\frac{1}{2}} = 0,$$

$$\sin \frac{A}{2} \cdot \alpha^{\frac{1}{2}} + \cos \frac{B}{2} (-\beta)^{\frac{1}{2}} + \sin \frac{C}{2} \cdot \gamma^{\frac{1}{2}} = 0,$$

$$\sin \frac{A}{2} \cdot \alpha^{\frac{1}{2}} + \sin \frac{B}{2} \cdot \beta^{\frac{1}{2}} + \cos \frac{C}{2} (-\gamma)^{\frac{1}{2}} = 0.$$

We may remark that, at every point in the circle which touches BC externally, α is essentially negative, so that the form $(-\alpha)^{\frac{1}{2}}$ represents a real quantity. Similarly the appearance of $(-\beta)^{\frac{1}{2}}$, $(-\gamma)^{\frac{1}{2}}$ in the equations of the other two escribed circles may be accounted for*.

11. The next form of the general equation of the second degree which we propose to consider is that in which u', v', w', the respective coefficients of $2\beta\gamma$, $2\gamma\alpha$, $2\alpha\beta$, are all $= 0$. The equation then assumes the form

$$u\alpha^2 + v\beta^2 + w\gamma^2 = 0.$$

We observe in the first place, that if this equation represent a real conic, the coefficients of α^2, β^2, γ^2, cannot be all of the same sign. Suppose the coefficient of α^2 to be of a different sign from the other two, then writing, for convenience of future investigations, L^2, $-M^2$, $-N^2$ for u, v, w respectively, our equation assumes the form

$$L^2\alpha^2 - M^2\beta^2 - N^2\gamma^2 = 0.$$

12. We have now to enquire how this conic is related to the triangle of reference.

Putting $\beta = 0$, we get

$$L\alpha = \pm N\gamma.$$

* If these equations be rationalised, and the sines and cosines of $\frac{A}{2}$, $\frac{B}{2}$, $\frac{C}{2}$ be expressed in terms of the sides, they assume the following forms:

$$a^2(s-a)^2\alpha^2 + b^2(s-b)^2\beta^2 + c^2(s-c)^2\gamma^2 - 2bc(s-b)(s-c)\beta\gamma - 2ca(s-c)(s-a)\gamma\alpha$$
$$- 2ab(s-a)(s-b)\alpha\beta = 0,$$

$$a^2s^2\alpha^2 + b^2(s-c)^2\beta^2 + c^2(s-b)^2\gamma^2 - 2bc(s-b)(s-c)\beta\gamma + 2cas(s-b)\gamma\alpha$$
$$+ 2abs(s-c)\alpha\beta = 0,$$

The interpretation of this equation is, that the two straight lines drawn from B to the points in which the conic is cut by CA, form, with BC, BA, an harmonic pencil.

It may similarly be proved that the two straight lines drawn from C to the points in which the conic is cut by AB, form, with CA, CB, an harmonic pencil.

If we put $\alpha = 0$, we get

$$M\beta = \pm \sqrt{(-1)}\, N\gamma,$$

shewing that BC cuts the conic in two imaginary points. The analytical condition of harmonic section is, however, satisfied here also.

13. We may next investigate the equations of the tangents drawn through the points A, B, C.

If in the equation of the conic we put $L\alpha = N\gamma$, we get $\beta = 0$, shewing that the straight line $L\alpha - N\gamma = 0$ meets the conic in two coincident points, and, therefore, touches it.

Similarly

$$L\alpha + N\gamma = 0, \quad L\alpha - M\beta = 0, \quad L\alpha + M\beta = 0,$$

are tangents to the conic.

The tangents to the conic drawn through A would be analytically represented by the equations

$$M\beta = \sqrt{(-1)}\, N\gamma, \quad M\beta = -\sqrt{(-1)}\, N\gamma,$$

which shew that these tangents are *imaginary*, or that the point A lies within the concavity of the conic.

14. Since the two tangents drawn through B meet the conic in points situated in the line CA, it follows that CA is the chord of contact of tangents to the conic drawn through B, or that CA is the *polar* of B, and B the *pole* of CA with respect to the conic. Similarly, C, AB, stand to one another in the relation of pole and polar.

Again, since the pole of AB is the point C, and the pole of AC is the point B, it follows that the line joining B and C is the polar of the point of intersection of AB, AC, i.e. that A is the pole of BC, and BC the polar of A.

We come then to this conclusion, that when an equation of the second degree does not involve the terms $\beta\gamma$, $\gamma\alpha$, $\alpha\beta$, the conic represented by it is so related to the triangle of reference, that *each side of the triangle is the polar, with respect to the conic, of the opposite angular point**.

This is expressed by saying that the triangle is *self-conjugate* with respect to the conic; or that the three angular points of the triangle form *a conjugate triad*.

The geometrical properties of the conic having been thus established, we shall, in future investigations, write for the sake of symmetry of form, $-L^2$ instead of L^2, so that the equation of the conic will be written

$$L^2\alpha^2 + M^2\beta^2 + N^2\gamma^2 = 0.$$

It must here be borne in mind that one of the three quantities L, M, N is essentially *imaginary*.

15. Any two conic sections represented by such equations as

$$L^2\alpha^2 + M^2\beta^2 + N^2\gamma^2 = 0,$$

$$L'^2\alpha^2 + M'^2\beta^2 + N'^2\gamma^2 = 0,$$

have important relations to one another, which we proceed to consider.

They will of course intersect in four points, which may be real or imaginary. We will first suppose them real, and represent them by the letters P, Q, R, S.

Now the locus of the equation

$$(L^2M'^2 - L'^2M^2)\beta^2 + (L^2N'^2 - L'^2N^2)\gamma^2 = 0$$

passes through all the points P, Q, R, S; and, since it may be resolved into linear factors, represents two straight lines.

---

* If the coefficients of β^2 and γ^2 be equal, and the triangle of reference be right-angled at A, the form of the equation shews that A will be a focus of the conic, and BC the corresponding directrix.

Suppose them to be PQ and RS. The intersection of these two straight lines is given by the equations

$$(L^2M'^2 - L'^2M^2)^{\frac{1}{2}} \beta = (L'^2N^2 - L^2N'^2)^{\frac{1}{2}} \gamma,$$

$$(L^2M'^2 - L'^2M^2)^{\frac{1}{2}} \beta = - (L'^2N^2 - L^2N'^2)^{\frac{1}{2}} \gamma,$$

which evidently gives $\beta = 0$, $\gamma = 0$.

Hence PQ, RS intersect in A.

Similarly, PR, QS intersect in B, and PS, QR intersect in C. Hence, the angular points of the triangle of reference coincide with the intersections of the line joining each pair of points of intersection of the conics with the line joining the other pair. Hence also, if any number of conic sections be described about the same quadrangle*, and the diagonals of that quadrangle intersect in A, while the sides produced intersect in B and C, then A, B, C form, with respect to each of the circumscribing conics, a conjugate triad. The points A, B, C may themselves be called *vertices* of the quadrangle, or of the system of circumscribing conics.

It will be seen, from the preceding investigation, that *any* two conics which intersect in four real points can be reduced, by a proper choice of the triangle of reference, to the form

$$L^2\alpha^2 + M^2\beta^2 + N^2\gamma^2 = 0.$$

The same reduction may also be effected in every case with the reservation that if two of the points of intersection of the conics be real and two imaginary, then two of the angular points of the triangle of reference (or vertices) will be imaginary and the remaining one real. If all the points of intersection be imaginary, the vertices of the conics will be all real. This we shall prove hereafter.

16. *To find the condition that a given straight line may touch the conic.*

Let the equation of the straight line be

$$l\alpha + m\beta + n\gamma = 0.$$

---

* I employ the term *quadrangle* in preference to *quadrilateral*, considering a *quadrangle* as a figure primarily determined by four *points*, a *quadrilateral* by four indefinite *straight lines*.

CONDITION OF TANGENCY.

Where this meets the conic, we have
$$L^2(m\beta + n\gamma)^2 + l^2(M^2\beta^2 + N^2\gamma^2) = 0,$$
and, making the two values of $\beta : \gamma$ equal, we get
$$(L^2m^2 + M^2l^2)(L^2n^2 + N^2l^2) = L^4m^2n^2,$$
whence
$$M^2N^2l^2 + N^2L^2m^2 + L^2M^2n^2 = 0,$$
or
$$\frac{l^2}{L^2} + \frac{m^2}{M^2} + \frac{n^2}{N^2} = 0,$$
the required condition.

17. *To find the condition that the conic may be a parabola.*

Since every parabola satisfies the analytical condition of touching the line
$$a\alpha + b\beta + c\gamma = 0,$$
the required condition becomes
$$\frac{a^2}{L^2} + \frac{b^2}{M^2} + \frac{c^2}{N^2} = 0.$$

18. *To find the co-ordinates of the centre.*

Let B_3, B_1 be the points in which the conic is cut by CA, then, if B_3, B_1 be bisected in Q, the line BQ will pass through the centre.

Now, let f_3, 0, h_3 be the co-ordinates of B_3,
$$f_1, 0, h_1 \dots\dots\dots\dots\dots\dots B_1.$$
Then those of Q are
$$\frac{f_3 + f_1}{2}, \; 0, \; \frac{h_3 + h_1}{2},$$
and the equation of BQ will be
$$\frac{\gamma}{h_3 + h_1} = \frac{\alpha}{f_3 + f_1}.$$

F.

Now f_2, f_1 are the values of α given by the equations
$$L^2\alpha^2 + M^2\beta^2 + N^2\gamma^2 = 0,$$
$$\beta = 0,$$
$$a\alpha + b\beta + c\gamma = 2\Delta,$$
which, eliminating β, γ, are equivalent to
$$L^2c^2\alpha^2 + N^2(a\alpha - 2\Delta)^2 = 0,$$
whence
$$f_2 + f_1 = \frac{4\Delta \cdot N^2 a}{L^2c^2 + N^2a^2}.$$
Similarly
$$h_2 + h_1 = \frac{4\Delta \cdot L^2 c}{N^2a^2 + L^2c^2}.$$
Hence the equation of BQ is
$$\frac{\gamma}{L^2c} = \frac{\alpha}{N^2a},$$
or
$$\frac{N^2\gamma}{c} = \frac{L^2\alpha}{a}.$$

This gives one straight line on which the centre lies. It may be similarly proved to lie on the straight line
$$\frac{L^2\alpha}{a} = \frac{M^2\beta}{b}.$$

Therefore the co-ordinates of the centre are given by the equations
$$\frac{L^2\alpha}{a} = \frac{M^2\beta}{b} = \frac{N^2\gamma}{c}.$$

Combining therewith
$$a\alpha + b\beta + c\gamma = 2\Delta,$$
we get for the co-ordinates of the centre

$$2\Delta \frac{\frac{a}{L^2}}{\frac{a^2}{L^2} + \frac{b^2}{M^2} + \frac{c^2}{N^2}}, \quad 2\Delta \frac{\frac{b}{M^2}}{\frac{a^2}{L^2} + \frac{b^2}{M^2} + \frac{c^2}{N^2}}, \quad 2\Delta \frac{\frac{c}{N^2}}{\frac{a^2}{L^2} + \frac{b^2}{M^2} + \frac{c^2}{N^2}}.$$

Each of these becomes infinite when the conic is a parabola, as manifestly ought to be the case.

19. *To find the equation of the circle with respect to which the triangle of reference is self-conjugate.*

It is a distinguishing property of the circle that the line joining the centre with any other point is perpendicular to the polar of that point. Hence the line

$$\frac{M^2\beta}{b} - \frac{N^2\gamma}{c} = 0,$$

which joins the centre with the point A, must be perpendicular to $\alpha = 0$. This gives (see Art. 5, p. 8)

$$\frac{M^2}{b \cos B} = \frac{N^2}{c \cos C}.$$

Similarly, since the lines joining the centre with B, C are respectively perpendicular to

$$\beta = 0, \ \gamma = 0,$$

we shall have

$$\frac{N^2}{c \cos C} = \frac{L^2}{a \cos A}, \quad \frac{L^2}{a \cos A} = \frac{M^2}{b \cos B}.$$

Hence the equation of the required circle is

$$a \cos A \cdot \alpha^2 + b \cos B \cdot \beta^2 + c \cos C \cdot \gamma^2 = 0,$$

or $\quad \sin 2A \cdot \alpha^2 + \sin 2B \cdot \beta^2 + \sin 2C \cdot \gamma^2 = 0.$

It will be remarked that this circle will be *imaginary* unless one of the quantities $\sin 2A$, $\sin 2B$, $\sin 2C$ be negative, that is, unless one of the angles $2A$, $2B$, $2C$ be greater than two right angles, or *unless the triangle of reference be obtuse-angled*.

COR. By referring to the expressions for the co-ordinates of the centre of the conic, given in Art. 18, we see that at the centre of the circle we have

$$\alpha \cos A = \beta \cos B = \gamma \cos C.$$

Or, *the centre of the circle, with respect to which the triangle of reference is self-conjugate, coincides with the intersection of the perpendiculars drawn from the angular points to the opposite sides.* This is otherwise evident from geometrical considerations.

20. *To find the equation of the conic which touches two sides of the triangle of reference in the points where they meet the third.*

Let AB, AC be the two sides which the required conic touches in the points B, C. We then require that the constants in the equation

$$L\alpha^2 + M\beta^2 + N\gamma^2 + 2\lambda\beta\gamma + 2\mu\gamma\alpha + 2\nu\alpha\beta = 0$$

should be so related to one another, that when $\beta = 0$ we have the two values of $\alpha = 0$, and also when $\gamma = 0$ the two values of α may each $= 0$.

Hence the two equations

$$L\alpha^2 + N\gamma^2 + 2\mu\gamma\alpha = 0,$$

$$L\alpha^2 + M\beta^2 + 2\nu\alpha\beta = 0,$$

must both be identically satisfied when $\alpha = 0$, *and by no other value.* This requires that

$$N = 0, \quad \mu = 0, \quad M = 0, \quad \nu = 0.$$

Hence the equation reduces to

$$L\alpha^2 + 2\lambda\beta\gamma = 0,$$

or, writing $-k^2$ for $\dfrac{L}{2\lambda}$,

$$k^2\alpha^2 = \beta\gamma.$$

This equation, it will be observed, involves only *one* arbitrary constant, as ought to be the case, since when a tangent and its point of contact are given, the conic is thus subjected to two conditions, and, therefore, when two tangents and their points of contact are given, to four.

21. If any straight line whatever be drawn through A, and meet the conic in P, Q, and be represented by the equation
$$\beta = n^2\gamma,$$
then BP, BQ will be represented by the equations
$$k\alpha = n\gamma, \quad k\alpha = -n\gamma,$$
from the form of which it is apparent that BA, BP, BC, BQ form an harmonic pencil. Or *any chord of a conic is divided harmonically by the conic itself, any point on the chord, and the polar of the point with respect to the conic.*

22. We may observe, that the two straight lines represented by the equations
$$k\alpha = \omega\beta, \quad k\alpha = \frac{1}{\omega}\gamma,$$
intersect on this conic whatever be the value of ω. Hence any point on the conic may be expressed by giving the value of the ratio
$$\frac{k\alpha}{\beta}, \text{ or } \frac{\gamma}{k\alpha}.$$

If ω be the value of this ratio at any point, that point may be denoted by the letter ω*. The line joining the two points ω, ω' may be called the line $\omega\omega'$.

23. *To find the equation of the line $\omega\omega'$.*

Let the required equation be
$$k\alpha + m\beta + n\gamma = 0,$$
we have then to determine m and n.

Since, when $k\alpha = \omega\beta, \quad k\alpha = \frac{1}{\omega}\gamma,$

we get $1 + \frac{m}{\omega} + n\omega = 0.$

---

* This mode of expression is given by Salmon in his *Conic Sections*.

Similarly
$$1 + \frac{m}{\omega'} + n\omega' = 0.$$

Hence
$$m = \frac{\omega' - \omega}{\dfrac{\omega}{\omega'} - \dfrac{\omega}{\omega}} = -\frac{\omega\omega'}{\omega + \omega'},$$

and
$$n = -\frac{1}{\omega + \omega'}.$$

Hence the line $\omega\omega'$ is represented by the equation
$$\omega\omega'\beta + \gamma = (\omega + \omega')k\alpha.$$

24. *To find the equation of the tangent at ω.*

This is obtained at once, from the result of the preceding article, by simply putting $\omega' = \omega$. It will then be seen to be
$$\omega^2\beta + \gamma = 2\omega \cdot k\alpha.$$

25. *To find the pole of $\omega\omega'$.*

The pole of $\omega\omega'$ is the point of intersection of the tangents at ω, ω'.

It is therefore given by the equations
$$2\omega \cdot k\alpha - \omega^2\beta - \gamma = 0,$$
$$2\omega' \cdot k\alpha - \omega'^2\beta - \gamma = 0,$$

whence
$$\frac{k\alpha}{\omega^2 - \omega'^2} = \frac{\beta}{2(\omega - \omega')} = \frac{\gamma}{2\omega\omega'(\omega - \omega')},$$

or
$$\frac{2k\alpha}{\omega + \omega'} = \beta = \frac{\gamma}{\omega\omega'}.$$

26. *To find the condition that a given straight line may touch the conic.*

Let the equation of the given straight line be
$$l\alpha + m\beta + n\gamma = 0.$$

CONDITION OF TANGENCY.

From Art. 24, it appears that if this straight line touch the conic, it must admit of being put in the form

$$-2\omega k\alpha + \omega^2\beta + \gamma = 0;$$

$$\therefore -\frac{2\omega k}{l} = \frac{\omega^2}{m} = \frac{1}{n},$$

whence
$$l^2 = 4k^2 mn,$$
the required condition.

The co-ordinates of the point of contact of this line will be determined by the equations

$$-l\alpha = 2m\beta = 2n\gamma.$$

COR. By writing a, b, c respectively for l, m, n in the condition of tangency just investigated, we see that the necessary condition in order that the conic may be a parabola is

$$a^2 = 4k^2 bc.$$

Or *the equation of the parabola touching* AB, AC *in* B, C, *is*

$$a^2\alpha^2 = 4bc\beta\gamma.$$

27. *To find the centre of the conic.*

Since the conic touches AB, AC in BC, it follows that the straight line drawn through A, and the middle point of BC, will pass through the centre. The equation of this straight line is

$$b\beta - c\gamma = 0.$$

If f_1, g_1, h_1, f_2, g_2, h_2 be the co-ordinates of the points in which the straight line meets the conic, those of the centre will be

$$\frac{f_1+f_2}{2}, \frac{g_1+g_2}{2}, \frac{h_1+h_2}{2}.$$

Now f_1, f_2, g_2, g_1, h_1, h_2, are the respective values of α, β, γ, obtained from the equations

$$k^2\alpha^2 - \beta\gamma = 0,$$
$$b\beta - c\gamma = 0,$$
$$a\alpha + b\beta + c\gamma = 2\Delta.$$

Eliminating γ, α between these, we get
$$a^2b\beta^2 = 4k^2c(b\beta - \Delta)^2,$$
whence
$$\frac{g_1 + g_2}{2} = \frac{4k^2c\Delta}{4k^2bc - a^2}.$$
Similarly
$$\frac{h_1 + h_2}{2} = \frac{4k^2b\Delta}{4k^2bc - a^2}.$$

These are the values β, γ at the centre. The corresponding value of α may be ascertained by substitution in the equation
$$a\alpha + b\beta + c\gamma = 2\Delta$$
to be
$$\frac{-2a\Delta}{4k^2bc - a^2}.$$

These values all become infinite when $4k^2bc = a^2$, as manifestly ought to be the case, since, as has been shewn in Art. 24, the conic is then a parabola.

Examples.

1. A triangle is inscribed in a conic; prove that the points, in which each side intersects the tangent at the opposite angle, lie in a straight line.

2. A triangle is described about a conic; prove that the straight lines, joining each angular point with the point of contact of the opposite side, intersect in a point.

3. Find the equations of the normals to the conic $\lambda\beta\gamma + \mu\gamma\alpha + \nu\alpha\beta = 0$, drawn at the angular points of the triangle of reference; and prove that they will intersect in a point if
$$\frac{\lambda}{a}(\mu^2 - \nu^2) + \frac{\mu}{b}(\nu^2 - \lambda^2) + \frac{\nu}{c}(\lambda^2 - \mu^2) = 0.$$

4. If the normals to a circumscribing conic, at the angular points of the triangle of reference, meet in a point, prove that the

locus of the centre of the conic is made up of the curve represented by the equation

$$\frac{a}{a}(\beta^2 - \gamma^2) + \frac{\beta}{b}(\gamma^2 - a^2) + \frac{\gamma}{c}(a^2 - \beta^2) = 0,$$

and of the three straight lines which join the middle points of the sides of the triangle of reference.

5. Three conics are drawn, touching respectively each pair of the sides of a triangle at the angular points where they meet the third side, and all intersecting in a point. Prove that the three tangents at their common point meet the sides of the triangle which intersect their respective conics in three points lying in a straight line; and that the other common tangents to each pair of conics intersect the sides of the triangle which touch the several pairs of conics in the same three points.

6. Prove that the points of intersection of the opposite sides of any quadrangle, and the point of intersection of the diagonals, form a conjugate triad with respect to any conic described about the quadrangle.

7. If R be the radius of the circle described about the triangle of reference, ρ that of the circle with respect to which the triangle of reference is self-conjugate, prove that

$$\rho^2 + 4R^2 \cos A \cos B \cos C = 0.$$

8. If BC, CA, AB be three given tangents to a conic, P, Q, R three points on the curve, and if the areas of the triangles PBC, PCA, PAB be denoted by p_1, p_2, p_3, respectively, and those of the triangles obtained by successively writing Q and R in place of P by q_1, q_2, q_3, r_1, r_2, r_3, prove that

$$(p_1 q_2 r_3)^{\frac{1}{2}} - (p_1 q_3 r_2)^{\frac{1}{2}} + (p_2 q_3 r_1)^{\frac{1}{2}} - (p_2 q_1 r_3)^{\frac{1}{2}} + (p_3 q_1 r_2)^{\frac{1}{2}} - (p_3 q_2 r_1)^{\frac{1}{2}} = 0.$$

9. Prove that the diagonals of any quadrilateral described about a conic, and the lines joining the points of contact of opposite sides, all intersect in a point.

10. A system of conics is described touching three straight lines; prove that, if one of the foci move along a given straight line, the other will describe a conic about a triangle.

Hence prove that the circle, which passes through the points of intersection of three tangents to a parabola, passes also through the focus.

11. Prove that the equation of the line passing through the feet of the perpendiculars from a point a_1, β_1, γ_1, of the circle $a\beta\gamma + b\gamma a + ca\beta = 0$, on the sides of the triangle of reference, may be put in the form,

$$\frac{c\beta_1 + b\gamma_1}{\beta_1 \cos C - \gamma_1 \cos B} a\alpha + \frac{a\gamma_1 + ca_1}{\gamma_1 \cos A - a_1 \cos C} b\beta + \frac{ba_1 + a\beta_1}{a_1 \cos B - \beta_1 \cos A} c\gamma = 0.$$

12. Shew that the axis of the parabola, whose equation is $a^2\alpha^2 = 4bc\beta\gamma$, is given by the equation

$$(c + b \cos A)\beta - (b + c \cos A)\gamma = \tfrac{1}{2}\left(\frac{b}{c} - \frac{c}{b}\right)a\alpha.$$

13. The equation of the directrix of the parabola, which touches the sides of the triangle of reference, and also the straight line $l\alpha + m\beta + n\gamma = 0$, is

$$a \cos A \left(\frac{1}{m} - \frac{1}{n}\right) + \beta \cos B \left(\frac{1}{n} - \frac{1}{l}\right) + \gamma \cos C \left(\frac{1}{l} - \frac{1}{m}\right).$$

14. If the equation

$$(l\alpha)^{\frac{1}{2}} + (m\beta)^{\frac{1}{2}} + (n\gamma)^{\frac{1}{2}} = 0$$

represent a parabola, the equation of its axis is

$$\frac{a^2\alpha}{l}\left(\frac{b^4}{m^2} - \frac{c^4}{n^2}\right) + \frac{b^2\beta}{m}\left(\frac{c^4}{n^2} - \frac{a^4}{l^2}\right) + \frac{c^2\gamma}{n}\left(\frac{a^4}{l^2} - \frac{b^4}{m^2}\right) = 0.$$

CHAPTER III.

ON ELIMINATION BETWEEN LINEAR EQUATIONS.

1. BEFORE entering upon the discussion of the conic represented by the general equation of the second degree, it will be necessary to devote a few pages to the subject of elimination between homogeneous linear equations, and to explain some of the terms recently introduced in connection with this branch of analysis.

We shall, however, only state and prove such elementary theorems as will be necessary in our future investigations; referring the reader who may be desirous of fuller information to Salmon's *Lessons on the Higher Algebra;* Spottiswoode, *On Determinants* (the second edition of which will be found in Crelle's *Journal*, t. 51, pp. 209, 328), and to the original memoirs communicated to various scientific Journals by Messrs Boole, Sylvester, Cayley, and others.

2. If we have given n homogeneous linear equations, connecting n unknown quantities $x_1, x_2 \ldots x_n$, such as

$$a_1 x_1 + a_2 x_2 + \ldots + a_n x_n = 0,$$
$$b_1 x_1 + b_2 x_2 + \ldots + b_n x_n = 0,$$
$$\vdots$$
$$k_1 x_1 + k_2 x_2 + \ldots + k_n x_n = 0,$$

the quantities $x_1, x_2 \ldots x_n$ can be eliminated between them, and the result of the elimination may be expressed by omitting $x_1, x_2 \ldots x_n$ and writing the coefficients only in the order in which they appear in the given equations, thus

$$\begin{vmatrix} a_1, a_2 \ldots a_n \\ b_1, b_2 \ldots b_n \\ \vdots \\ k_1, k_2 \ldots k_n \end{vmatrix} = 0.$$

The left-hand member of this equation is what is called the *determinant* of the given system of equations.

We proceed to investigate the law of its formation.

3. First, suppose we have two equations,
$$a_1 x_1 + a_2 x_2 = 0,$$
$$b_1 x_1 + b_2 x_2 = 0.$$

Multiply the first by b_2, the second by a_2, and subtract, and we get
$$a_1 b_2 - a_2 b_1 = 0.$$

Hence $\begin{vmatrix} a_1, a_2 \\ b_1, b_2 \end{vmatrix} = a_1 b_2 - a_2 b_1.$

We may remark in passing that we shall obtain the same result by eliminating λ_1, λ_2 between the equations
$$a_1 \lambda_1 + b_1 \lambda_2 = 0,$$
$$a_2 \lambda_1 + b_2 \lambda_2 = 0.$$

Hence $\begin{vmatrix} a_1, b_1 \\ a_2, b_2 \end{vmatrix} = \begin{vmatrix} a_1, a_2 \\ b_1, b_2 \end{vmatrix}.$

A like theorem will be proved to be true for all determinants.

4. Next, suppose we have the three equations
$$a_1 x_1 + a_2 x_2 + a_3 x_3 = 0,$$
$$b_1 x_1 + b_2 x_2 + b_3 x_3 = 0,$$
$$c_1 x_1 + c_2 x_2 + c_3 x_3 = 0.$$

Multiply these equations in order by the arbitrary multipliers $\lambda_1, \lambda_2, \lambda_3$, and add them together. Let the two ratios $\lambda_1 : \lambda_2 : \lambda_3$ be determined by the conditions that the coefficients of x_2 and x_3 in the resulting equation shall each be zero, i.e. let

$$\left. \begin{array}{l} a_2\lambda_1 + b_2\lambda_2 + c_2\lambda_3 = 0 \\ a_3\lambda_1 + b_3\lambda_2 + c_3\lambda_3 = 0 \end{array} \right\} \dots\dots\dots(A).$$

The resulting equation is then reduced to

$$(a_1\lambda_1 + b_1\lambda_2 + c_1\lambda_3)x_1 = 0,$$

which requires that

$$a_1\lambda_1 + b_1\lambda_2 + c_1\lambda_3 = 0 \dots\dots\dots(B).$$

Multiply the first of equations (A) by a_3, the second by a_2, and subtract, we then get

$$(a_3b_2 - a_2b_3)\lambda_2 + (c_2a_3 - c_3a_2)\lambda_3 = 0,$$

or

$$\frac{\lambda_2}{c_2a_3 - c_3a_2} = \frac{\lambda_3}{a_2b_3 - a_3b_2}$$

$$= \frac{\lambda_1}{b_2c_3 - b_3c_2}, \text{ by symmetry}\dots(C).$$

Hence, dividing each term of (B) by the corresponding member of (C) we get

$$a_1(b_2c_3 - b_3c_2) + b_1(c_2a_3 - c_3a_2) + c_1(a_2b_3 - a_3b_2),$$

or

$$\begin{vmatrix} a_1, & a_2, & a_3 \\ b_1, & b_2, & b_3 \\ c_1, & c_2, & c_3 \end{vmatrix} = a_1(b_2c_3 - b_3c_2) + b_1(c_2a_3 - c_3a_2) \\ + c_1(a_2b_3 - a_3b_2)$$

$$= a_1 \begin{vmatrix} b_2, & b_3 \\ c_2, & c_3 \end{vmatrix} + b_1 \begin{vmatrix} c_2, & c_3 \\ a_2, & a_3 \end{vmatrix} + c_1 \begin{vmatrix} a_2, & a_3 \\ b_2, & b_3 \end{vmatrix}$$

$$= a_1 \begin{vmatrix} b_2, & b_3 \\ c_2, & c_3 \end{vmatrix} - b_1 \begin{vmatrix} a_2, & a_3 \\ c_2, & c_3 \end{vmatrix} + c_1 \begin{vmatrix} a_2, & a_3 \\ b_2, & b_3 \end{vmatrix}$$

It will be seen that the above process is really equivalent to that of eliminating λ_1, λ_2, λ_3 between the equations (A) and (B). Hence

$$\begin{vmatrix} a_1, & a_2, & a_3 \\ b_1, & b_2, & b_3 \\ c_1, & c_2, & c_3 \end{vmatrix} = \begin{vmatrix} a_1, & b_1, & c_1 \\ a_2, & b_2, & c_2 \\ a_3, & b_3, & c_3 \end{vmatrix}$$

5. Next, let us have the four equations

$$a_1 x_1 + a_2 x_2 + a_3 x_3 + a_4 x_4 = 0,$$
$$b_1 x_1 + b_2 x_2 + b_3 x_3 + b_4 x_4 = 0,$$
$$c_1 x_1 + c_2 x_2 + c_3 x_3 + c_4 x_4 = 0,$$
$$d_1 x_1 + d_2 x_2 + d_3 x_3 + d_4 x_4 = 0.$$

To effect the elimination, multiply the equations in order by λ_1, λ_2, λ_3, λ_4, add them, and equate the coefficients of x_2, x_3, x_4 severally to zero. We shall then have

$$\left. \begin{array}{l} a_2 \lambda_1 + b_2 \lambda_2 + c_2 \lambda_3 + d_2 \lambda_4 = 0 \\ a_3 \lambda_1 + b_3 \lambda_2 + c_3 \lambda_3 + d_3 \lambda_4 = 0 \\ a_4 \lambda_1 + b_4 \lambda_2 + c_4 \lambda_3 + d_4 \lambda_4 = 0 \end{array} \right\} \quad \ldots\ldots\ldots \text{(A')},$$

which equations involve as a consequence

$$a_1 \lambda_1 + b_1 \lambda_2 + c_1 \lambda_3 + d_1 \lambda_4 = 0 \ldots\ldots\ldots \text{(B')}.$$

To determine the three ratios $\lambda_1 : \lambda_2 : \lambda_3 : \lambda_4$, multiply equations (A') in order by μ_2, μ_3, μ_4, add, and equate to zero the coefficients of λ_3, λ_4. We thus get

$$\left. \begin{array}{l} c_2 \mu_2 + c_3 \mu_3 + c_4 \mu_4 = 0 \\ d_2 \mu_2 + d_3 \mu_3 + d_4 \mu_4 = 0 \end{array} \right\} \quad \ldots\ldots\ldots\ldots \text{(C')}.$$

Also $(a_2 \mu_2 + a_3 \mu_3 + a_4 \mu_4) \lambda_1 + (b_2 \mu_2 + b_3 \mu_3 + b_4 \mu_4) \lambda_2 = 0$

whence $\dfrac{\lambda_1}{b_2 \mu_2 + b_3 \mu_3 + b_4 \mu_4} = \dfrac{\lambda_2}{-(a_2 \mu_2 + a_3 \mu_3 + a_4 \mu_4)}.$

Now, treating equations (C') as equations (A) were treated, we see that

$$\frac{\mu_2}{c_3 d_4 - c_4 d_3} = \frac{\mu_3}{c_4 d_2 - c_2 d_4} = \frac{\mu_4}{c_2 d_3 - c_3 d_2},$$

or

$$\frac{\mu_2}{\begin{vmatrix} c_3, & d_3 \\ c_4, & d_4 \end{vmatrix}} = \frac{\mu_3}{\begin{vmatrix} c_4, & d_4 \\ c_2, & d_2 \end{vmatrix}} = \frac{\mu_4}{\begin{vmatrix} c_2, & d_2 \\ c_3, & d_3 \end{vmatrix}}$$

whence

$$\frac{\lambda_1}{b_2 \begin{vmatrix} c_3, & d_3 \\ c_4, & d_4 \end{vmatrix} + b_3 \begin{vmatrix} c_4, & d_4 \\ c_2, & d_2 \end{vmatrix} + b_4 \begin{vmatrix} c_2, & d_2 \\ c_3, & d_3 \end{vmatrix}}$$

$$= - \frac{\lambda_2}{a_2 \begin{vmatrix} c_3, & d_3 \\ c_4, & d_4 \end{vmatrix} + a_3 \begin{vmatrix} c_4, & d_4 \\ c_2, & d_2 \end{vmatrix} + a_4 \begin{vmatrix} c_2, & d_2 \\ c_3, & d_3 \end{vmatrix}}$$

or

$$\frac{\lambda_1}{\begin{vmatrix} b_2, & c_2, & d_2 \\ b_3, & c_3, & d_3 \\ b_4, & c_4, & d_4 \end{vmatrix}} = - \frac{\lambda_2}{\begin{vmatrix} a_2, & c_2, & d_2 \\ a_3, & c_3, & d_3 \\ a_4, & c_4, & d_4 \end{vmatrix}}$$

which, by symmetry, are equal to

$$\frac{\lambda_3}{\begin{vmatrix} a_2, & b_2, & d_2 \\ a_3, & b_3, & d_3 \\ a_4, & b_4, & d_4 \end{vmatrix}} = - \frac{\lambda_4}{\begin{vmatrix} a_2, & b_2, & c_2 \\ a_3, & b_3, & c_3 \\ a_4, & b_4, & c_4 \end{vmatrix}}$$

These equations may be more conveniently written in the following equivalent forms:

$$\frac{\lambda_1}{\begin{vmatrix} b_2, & b_3, & b_4 \\ c_2, & c_3, & c_4 \\ d_2, & d_3, & d_4 \end{vmatrix}} = - \frac{\lambda_2}{\begin{vmatrix} a_2, & a_3, & a_4 \\ c_2, & c_3, & c_4 \\ d_2, & d_3, & d_4 \end{vmatrix}} = \frac{\lambda_3}{\begin{vmatrix} a_2, & a_3, & a_4 \\ b_2, & b_3, & b_4 \\ d_2, & d_3, & d_4 \end{vmatrix}} = - \frac{\lambda_4}{\begin{vmatrix} a_2, & a_3, & a_4 \\ b_2, & b_3, & b_4 \\ c_2, & c_3, & c_4 \end{vmatrix}}$$

Eliminating by means of these equations λ_1, λ_2, λ_3, λ_4, from equation B', we get, as the result of the elimination of x_1, x_2, x_3, x_4 between the four given equations,

64   TRILINEAR CO-ORDINATES.

$$\begin{vmatrix} a_1, & a_2, & a_3, & a_4 \\ b_1, & b_2, & b_3, & b_4 \\ c_1, & c_2, & c_3, & c_4 \\ d_1, & d_2, & d_3, & d_4 \end{vmatrix} = a_1 \begin{vmatrix} b_2, & b_3, & b_4 \\ c_2, & c_3, & c_4 \\ d_2, & d_3, & d_4 \end{vmatrix} - b_1 \begin{vmatrix} a_2, & a_3, & a_4 \\ c_2, & c_3, & c_4 \\ d_2, & d_3, & d_4 \end{vmatrix}$$

$$+ c_1 \begin{vmatrix} a_2, & a_3, & a_4 \\ b_2, & b_3, & b_4 \\ d_2, & d_3, & d_4 \end{vmatrix} - d_1 \begin{vmatrix} a_2, & a_3, & a_4 \\ b_2, & b_3, & b_4 \\ d_2, & d_3, & d_4 \end{vmatrix}$$

And since the above process is equivalent to the elimination of $\lambda_1, \lambda_2, \lambda_3, \lambda_4$, between the equations (A') and (B'), we see that

$$\begin{vmatrix} a_1, & a_2, & a_3, & a_4 \\ b_1, & b_2, & b_3, & b_4 \\ c_1, & c_2, & c_3, & c_4 \\ d_1, & d_2, & d_3, & d_4 \end{vmatrix} = \begin{vmatrix} a_1, & b_1, & c_1, & d_1 \\ a_2, & b_2, & c_2, & d_2 \\ a_3, & b_3, & c_3, & d_3 \\ a_4, & b_4, & c_4, & d_4 \end{vmatrix}$$

6. The law of formation will be sufficiently obvious from the above investigations. If we have n horizontal and vertical rows, it may be similarly proved that

$$\begin{vmatrix} a_1, & a_2, & a_3 \ldots a_n \\ b_1, & b_2, & b_3 \ldots b_n \\ c_1, & c_2, & c_3 \ldots c_n \\ \vdots & \vdots & \vdots \\ k_1, & k_2, & k_3 \ldots k_n \end{vmatrix} = \begin{vmatrix} a_1, & b_1, & c_1 \ldots k_1 \\ a_2, & b_2, & c_2 \ldots k_2 \\ a_3, & b_3, & c_3 \ldots k_3 \\ \vdots & \vdots & \vdots \\ a_n, & b_n, & c_n \ldots k_n \end{vmatrix}$$

$$= a_1 \begin{vmatrix} b_2, & b_3 \ldots b_n \\ c_2, & c_3 \ldots c_n \\ \vdots & \vdots \\ k_2, & k_3 \ldots k_n \end{vmatrix} - b_1 \begin{vmatrix} a_2, & a_3 \ldots a_n \\ c_2, & c_3 \ldots c_n \\ \vdots & \vdots \\ k_2, & k_3 \ldots k_n \end{vmatrix}$$

$$+ c_1 \begin{vmatrix} a_2, & a_3 \ldots a_n \\ b_2, & b_3 \ldots b_n \\ \vdots & \vdots \\ k_2, & k_3 \ldots k_n \end{vmatrix} + \ldots + (-1)^{n-1} k_1 \begin{vmatrix} a_2, & a_3 \ldots \\ b_2, & b_3 \ldots \\ \vdots & \vdots \\ \cdots \cdots \cdots \end{vmatrix}$$

It may also be proved that, if we have $n-1$ equations connecting n quantities $\lambda_1, \lambda_2 \ldots \lambda_n$, such as

$$a_2\lambda_1 + b_2\lambda_2 + c_2\lambda_3 + \ldots + k_2\lambda_n = 0,$$
$$a_3\lambda_1 + b_3\lambda_2 + c_3\lambda_3 + \ldots + k_3\lambda_n = 0,$$
$$a_4\lambda_1 + b_4\lambda_2 + c_4\lambda_3 + \ldots + k_4\lambda_n = 0,$$
$$\vdots$$
$$a_n\lambda_1 + b_n\lambda_2 + c_n\lambda_3 + \ldots + k_n\lambda_n = 0,$$

we shall obtain the following ratios between $\lambda_1, \lambda_2, \lambda_3, \ldots \lambda_n,$

$$\frac{\lambda_1}{\begin{vmatrix} b_2, c_2 \ldots k_2 \\ b_3, c_3 \ldots k_3 \\ \vdots \\ b_n, c_n \ldots k_n \end{vmatrix}} = \frac{\lambda_2}{-\begin{vmatrix} a_2, c_2 \ldots k_2 \\ a_3, c_3 \ldots k_3 \\ \vdots \\ a_n, c_n \ldots k_n \end{vmatrix}} = \frac{\lambda_3}{\begin{vmatrix} a_2, b_2 \ldots k_2 \\ a_3, b_3 \ldots k_3 \\ \vdots \\ a_n, b_n \ldots k_n \end{vmatrix}}$$

$$= \ldots = \frac{\lambda_n}{(-1)^{n-1}\begin{vmatrix} a_2, b_2 \ldots \ldots \\ a_3, b_3 \ldots \ldots \\ \vdots \quad \vdots \\ \ldots \ldots \ldots \end{vmatrix}}$$

By reference to the expanded values of the determinants

$$\begin{vmatrix} a_1, a_2 \\ b_1, b_2 \end{vmatrix} \quad \begin{vmatrix} a_1, a_2, a_3 \\ b_1, b_2, b_3 \\ c_1, c_2, c_3 \end{vmatrix}$$

it will be seen that the former contains $1 \cdot 2$ or two terms, the latter $1 \cdot 2 \cdot 3$ or six. It may also be proved that, if n quantities be eliminated from n linear homogeneous equations, the resulting determinant will contain $1 \cdot 2 \cdot 3 \ldots n$ terms. For, referring to the relation between determinants of n and $n-1$ rows, given in Arts. (4), (5), (6), it will be seen that this theorem is true for a determinant of n rows, if it be true for one of $n-1$. But it is true for three rows, therefore it is universally true.

7. The horizontal rows of a determinant are commonly spoken of as "lines," the vertical ones as "columns." It will

F.

be observed, moreover, that each term is the product of n factors, one taken from each line and from each column, and that the coefficients of one half of the terms are $+1$, of the other -1. To determine the sign of any particular term we proceed as follows. Considering for simplicity the case of three rows, we have

$$\begin{vmatrix} a_1, a_2, a_3 \\ b_1, b_2, b_3 \\ c_1, c_2, c_3 \end{vmatrix} = a_1 b_2 c_3 - a_1 b_3 c_2 + a_2 b_3 c_1 - a_2 b_1 c_3 + a_3 b_1 c_2 - a_3 b_2 c_1.$$

Here we observe, first, that (the factors of each term being arranged in alphabetical order, that is, in the order of the columns) the term $a_1 b_2 c_3$ (in which the suffixes follow the arithmetical order, that is, the order of the lines) has a positive coefficient. Now every other term may be formed from this by making each suffix change places with either of its adjacent suffixes a sufficient number of times. Thus the term $a_1 b_3 c_2$ is produced by simply making the suffixes 2 and 3 exchange places. The term $a_3 b_1 c_2$ is produced by making the suffix 3 change places, first with 2, and next with 1, which is then adjacent to it. If this process of interchanging the suffixes of two *consecutive* letters be called a "permutation," we may enunciate the following law, which by inspection will be seen to hold.

"Every term derived from the first by an odd number of permutations has a negative sign. Every term formed by an even number of permutations has a positive sign."

Thus, it will be observed that the terms $a_1 b_3 c_2$, $a_2 b_1 c_3$, each of which is derived from $a_1 b_2 c_3$ by one permutation, have negative signs. The terms $a_2 b_3 c_1$, $a_3 b_1 c_2$, each formed by two permutations, have positive signs. The term $a_3 b_2 c_1$, formed by three permutations, has a negative sign.

In like manner, in the case of a determinant of four rows, if $a_1 b_2 c_3 d_4$ have a positive sign, such a term as $a_2 b_1 c_4 d_3$, derived by two permutations, will have a positive sign, while $a_4 b_1 c_2 d_3$, derived by three, has a negative sign.

8. *The sign of a determinant is changed by interchanging any two consecutive lines or columns.*

CHANGE OF SIGN.

In the first place, we observe that

$$\begin{vmatrix} a_1, & a_2 \\ b_1, & b_2 \end{vmatrix} = a_1 b_2 - a_2 b_1 = -(b_1 a_2 - b_2 a_1) = - \begin{vmatrix} b_1, & b_2 \\ a_1, & a_2 \end{vmatrix}$$

Again,

$$\begin{vmatrix} a_1, & a_2, & a_3 \\ b_1, & b_2, & b_3 \\ c_1, & c_2, & c_3 \end{vmatrix} = a_1 \begin{vmatrix} b_2, & b_3 \\ c_2, & c_3 \end{vmatrix} - b_1 \begin{vmatrix} a_2, & a_3 \\ c_2, & c_3 \end{vmatrix} + c_1 \begin{vmatrix} a_2, & a_3 \\ b_2, & b_3 \end{vmatrix}$$

$$= - a_1 \begin{vmatrix} c_2, & c_3 \\ b_2, & b_3 \end{vmatrix} + c_1 \begin{vmatrix} a_2, & a_3 \\ b_2, & b_3 \end{vmatrix} - b_1 \begin{vmatrix} a_2, & a_3 \\ c_2, & c_3 \end{vmatrix}$$

by what has been shewn above,

$$= - \begin{vmatrix} a_1, & a_2, & a_3 \\ c_1, & c_2, & c_3 \\ b_1, & b_2, & b_3 \end{vmatrix}$$

The theorem enunciated is thus proved for determinants of two and of three rows, and may by successive inductions be extended to any number.

COR. It hence follows that, if any two lines or columns of a determinant be identical, the determinant will vanish. For we see, by the theorem, that

$$\begin{vmatrix} a_1, & a_2, & a_3 \\ b_1, & b_2, & b_3 \\ b_1, & b_2, & b_3 \end{vmatrix} = - \begin{vmatrix} a_1, & a_2, & a_3 \\ b_1, & b_2, & b_3 \\ b_1, & b_2, & b_3 \end{vmatrix}$$

and therefore $= 0$.

9. We see that

$$\begin{vmatrix} ma_1, & a_2, & a_3 \\ mb_1, & b_2, & b_3 \\ mc_1, & c_2, & c_3 \end{vmatrix} = ma_1 \begin{vmatrix} b_2, & b_3 \\ c_2, & c_3 \end{vmatrix} - mb_1 \begin{vmatrix} a_2, & a_3 \\ c_2, & c_3 \end{vmatrix}$$

$$+ mc_1 \begin{vmatrix} a_2, & a_3 \\ b_2, & b_3 \end{vmatrix}$$

$$= m \begin{vmatrix} a_1, & a_2, & a_3 \\ b_1, & b_2, & b_3 \\ c_1, & c_2, & c_3 \end{vmatrix}$$

Hence, *if all the terms in any line or column of a determinant be multiplied by any given quantity, the determinant itself will be multiplied by the same quantity.*

10. DEF. From any given determinant, other determinants may be formed, by omitting an equal number of lines and columns of the given determinants. These are termed MINORS of the given determinant, and are called first, second, &c. minors, according as one, two, &c. lines and columns have been omitted. Thus

$$\begin{vmatrix} b_2, & b_3 \\ c_2, & c_3 \end{vmatrix} \qquad \begin{vmatrix} a_1, & a_3 \\ b_1, & b_3 \end{vmatrix}$$

are first minors of

$$\begin{vmatrix} a_1, & a_2, & a_3 \\ b_1, & b_2, & b_3 \\ c_1, & c_2, & c_3 \end{vmatrix}$$

11. *To investigate the relation which must hold among the coefficients* L, M, N, λ, μ, ν, *in order that the quadratic function*

$$L\alpha^2 + M\beta^2 + N\gamma^2 + 2\lambda\beta\gamma + 2\mu\gamma\alpha + 2\nu\alpha\beta$$

may be the product of two factors of the first degree in α, β, γ.

Let $p\alpha + q\beta + r\gamma$ be one factor, then $\dfrac{L}{p}\alpha + \dfrac{M}{q}\beta + \dfrac{N}{r}\gamma$ must be the other. Hence

$$(p\alpha + q\beta + r\gamma)\left(\frac{L}{p}\alpha + \frac{M}{q}\beta + \frac{N}{r}\gamma\right) = L\alpha^2 + M\beta^2 + N\gamma^2$$
$$+ 2\lambda\beta\gamma + 2\mu\gamma\alpha + 2\nu\alpha\beta, \text{ identically,}$$

therefore
$$M\frac{r}{q} + N\frac{q}{r} = 2\lambda,$$

$$N\frac{p}{r} + L\frac{r}{p} = 2\mu,$$

$$L\frac{q}{p} + M\frac{p}{q} = 2\nu,$$

DISCRIMINANT.

whence
$$L\frac{qr}{p} = \mu q + \nu r - \lambda p,$$

$$M\frac{rp}{q} = \nu r + \lambda p - \mu q,$$

$$N\frac{pq}{r} = \lambda p + \mu q - \nu r,$$

$$\therefore LMN \cdot pqr = (\mu q + \nu r - \lambda p)(\nu r + \lambda p - \mu q)(\lambda p + \mu q - \nu r),$$
$$LML\lambda^2 \cdot pqr = \lambda^2 p^2 (\mu q + \nu r - \lambda p),$$
$$M\mu^2 pqr = \mu^2 q^2 (\nu r + \lambda p - \mu q),$$
$$N\nu^2 pqr = \nu^2 r^2 (\lambda p + \mu q - \nu r),$$
$$\therefore pqr(LMN - L\lambda^2 - M\mu^2 - N\nu^2) = -2\lambda p \cdot \mu q \cdot \nu r,$$
$$\therefore LMN - L\lambda^2 - M\mu^2 - N\nu^2 + 2\lambda\mu\nu = 0,$$

which may also be written
$$\begin{vmatrix} L, & \nu, & \mu \\ \nu, & M, & \lambda \\ \mu, & \lambda, & N \end{vmatrix} = 0.$$

This expression
$$\begin{vmatrix} L, & \nu, & \mu \\ \nu, & M, & \lambda \\ \mu, & \lambda, & N \end{vmatrix}$$

the evanescence of which is the necessary condition that the given quadratic function may break up into two factors, is termed the *Discriminant* of that function.

12. PASCAL'S THEOREM.

From the analytical result stated in Art. 6 of the present chapter, that the value of a determinant is not altered by changing its lines into columns and its columns into lines, we obtain a proof of Pascal's theorem, which asserts that

If a hexagon be inscribed in a conic, and the pairs of opposite sides be produced to intersect, the points of intersection lie in the same straight line.

Let $AFBDCE$ be the conic; take ABC as the triangle of reference, and let the equation of the conic be

$$\frac{\lambda}{\alpha} + \frac{\mu}{\beta} + \frac{\nu}{\gamma} = 0 \quad \ldots\ldots\ldots\ldots\ldots\ldots (1).$$

Let the equation of AE be $\beta = n_2\gamma$, of AF be $\gamma = m_3\beta$,

$BF \ldots \gamma = l_3\alpha$, of $BD \ldots \alpha = n_1\gamma$,

$CD \ldots \alpha = m_1\beta$, of $CE \ldots \beta = l_2\alpha$.

Then, since D lies in the conic (1), we have $\lambda + \mu m_1 + \nu n_1 = 0$,

E .... $\lambda l_2 + \mu + \nu n_2 = 0$,

F .... $\lambda l_3 + \mu m_3 + \nu = 0$,

whence $\begin{vmatrix} 1, & m, & n_1 \\ l_2, & 1, & n_2 \\ l_3, & m_3, & 1 \end{vmatrix} = 0 \ldots\ldots\ldots\ldots\ldots\ldots (2)$

is the necessary condition that the six points A, F, B, D, C, E may lie in a conic.

Again, if the pairs of opposite sides intersect in points lying in a straight line, let the equation of that straight line be $p\alpha + q\beta + r\gamma = 0$. Then, since

BF and CE intersect in this line, we have $\quad p + ql_2 + \nu l_3 = 0$,

CD and AF .... $pm_1 + q + \nu m_3 = 0$,

AE and BD .... $pn_1 + qn_2 + \nu = 0$,

whence $\begin{vmatrix} 1, & l_2, & l_3 \\ m_1, & 1, & m_3 \\ n_1, & n_2, & 1 \end{vmatrix} = 0 \ldots\ldots\ldots\ldots\ldots\ldots (3)$

as the condition that these points of intersection may lie in the same straight line. But (2) and (3) are identical. Hence the proposition is proved.

13. From Pascal's Theorem many interesting consequences may be deduced. Thus, if the point F coincide with A, D with B, E with C, then AF, BD, CE become the tangents at A, B, C, respectively, and we obtain the theorem enunciated in Ex. 1, Chap. II. Again, by supposing D to

PASCAL'S THEOREM.

coincide with B, and E with C, we readily obtain the following theorem: "If the opposite sides of a quadrilateral, inscribed in a conic, be produced to meet, and likewise the pairs of tangents at opposite angles of the quadrilateral, the four points of intersection will lie in the same straight line."

And, by supposing F to coincide with A, we obtain a geometrical construction, by which, having given five points of a conic, we can draw a tangent at any one of them. For, since AF then becomes the tangent at A, we see that, if AE, DB be produced to meet in G, AB, EC in H, and GH intersect CD in I, then AI will be the tangent at A.

EXAMPLES.

1. Prove that

$$\begin{vmatrix} a, & b, & c, & d \\ b, & a, & d, & c \\ c, & d, & a, & b \\ d, & c, & b, & a \end{vmatrix} = (a+b+c+d)(a-b+c-d)(a-b-c+d)(a+b-c-d).$$

2. If $\begin{vmatrix} b_2, & b_3 \\ c_2, & c_3 \end{vmatrix} = A_1$, $\begin{vmatrix} b_3, & b_1 \\ c_3, & b_1 \end{vmatrix} = A_2, \ldots$, prove that

$$\begin{vmatrix} B_2, & B_3 \\ C_2, & C_3 \end{vmatrix} = a_1 \begin{vmatrix} a_1, & a_2, & a_3 \\ b_1, & b_2, & b_3 \\ c_1, & c_2, & c_3 \end{vmatrix}$$

and that

$$\begin{vmatrix} A_1, & A_2, & A_3 \\ B_1, & B_2, & B_3 \\ C_1, & C_2, & C_3 \end{vmatrix} = \begin{vmatrix} a_1, & a_2, & a_3 \\ b_1, & b_2, & b_3 \\ c_1, & c_2, & c_3 \end{vmatrix}^2.$$

3. If $\begin{vmatrix} b, & c \\ y, & z \end{vmatrix} = A$, $\begin{vmatrix} c, & a \\ z, & x \end{vmatrix} = B$, $\begin{vmatrix} a, & b \\ x, & y \end{vmatrix} = C$,

and $\begin{vmatrix} b', & c' \\ y, & z \end{vmatrix} = A'$, $\begin{vmatrix} c', & a' \\ z, & x \end{vmatrix} = B'$, $\begin{vmatrix} a', & b' \\ x, & y \end{vmatrix} = C'$,

prove that $\begin{vmatrix} B, & C \\ B', & C' \end{vmatrix}^2 + \begin{vmatrix} C, & A \\ C', & A' \end{vmatrix}^2 + \begin{vmatrix} A, & B \\ A', & B' \end{vmatrix}^2 = \begin{vmatrix} a, & b, & c \\ a', & b', & c' \\ x, & y, & z \end{vmatrix}^2 (x^2 + y^2 + z^2).$

4. Prove that $\begin{vmatrix} 0, & 1, & 1, & 1, & \ldots \\ 1, & 0, & a+b, & a+c \ldots \\ 1, & b+a, & 0, & b+c \ldots \\ 1, & c+a, & c+b, & 0, & \ldots \\ \multicolumn{5}{c}{\dotfill} \end{vmatrix} = abc \ldots \left(\dfrac{1}{a} + \dfrac{1}{b} + \dfrac{1}{c} + \ldots \right).$

5. Prove that $\begin{vmatrix} m+n-y+z, & -y+z-l, & -y+z-l \\ -z+x-m, & n+l-z+x, & -z+x-m \\ -x+y-n, & -x+y-n, & l+m-x+y \end{vmatrix} = 0.$

6. Prove that $\begin{vmatrix} \dfrac{a}{b+c}, & \dfrac{b+c}{a}, & \dfrac{b+c}{a} \\ \dfrac{c+a}{b}, & \dfrac{b}{c+a}, & \dfrac{c+a}{b} \\ \dfrac{a+b}{c}, & \dfrac{a+b}{c}, & \dfrac{c}{a+b} \end{vmatrix} = \dfrac{2(a+b+c)^3}{(b+c)(c+a)(a+b)}.$

7. Prove that, if $\begin{vmatrix} A, & c, & b, & a \\ c, & B, & a, & \beta \\ b, & a, & C, & \gamma \\ a, & b, & \gamma, & F \end{vmatrix} = 0,$

then $\begin{vmatrix} A, & c, & b \\ c, & B, & a \\ b, & c, & C \end{vmatrix} \begin{vmatrix} A, & c, & a \\ c, & B, & \beta \\ a, & \beta, & F \end{vmatrix} = \begin{vmatrix} a, & A, & c \\ \beta, & c, & B \\ \gamma, & b, & a \end{vmatrix}^2.$

8. Prove that $\begin{vmatrix} (m+n)^2, & n^2, & m^2 \\ n^2, & (n+l)^2, & l^2 \\ l^2, & m^2, & (l+m)^2 \end{vmatrix} = 2(mn+nl+lm)^2.$

9. Prove that

$2 \begin{vmatrix} a^6, & a^3, & a^2, & a, & 1 \\ \beta^6, & \beta^3, & \beta^2, & \beta, & 1 \\ \multicolumn{5}{c}{\dotfill} \\ \epsilon^6, & \epsilon^3, & \epsilon^2, & \epsilon, & 1 \end{vmatrix} - 3 \begin{vmatrix} a^5, & a^4, & a^2, & a, & 1 \\ \beta^5, & \beta^4, & \beta^2, & \beta, & 1 \\ \multicolumn{5}{c}{\dotfill} \\ \epsilon^5, & \epsilon^4, & \epsilon^2, & \epsilon, & 1 \end{vmatrix}$

$= \dfrac{1}{2}(a-\beta)(a-\gamma)\ldots\{(a-\beta)^2 + (a-\gamma)^2 + \ldots\}.$

10. If
$$\frac{x}{a+\alpha}+\frac{y}{b+\alpha}+\frac{z}{c+\alpha}=1,$$

$$\frac{x}{a+\beta}+\frac{y}{b+\beta}+\frac{z}{c+\beta}=1,$$

$$\frac{x}{a+\gamma}+\frac{y}{b+\gamma}+\frac{z}{c+\gamma}=1,$$

find the values of x, y, z; and prove that

$$\frac{x}{a+\lambda}+\frac{y}{b+\lambda}+\frac{z}{c+\lambda}=1+\frac{(\alpha-\lambda)(\beta-\lambda)(\gamma-\lambda)}{(a+\lambda)(b+\lambda)(c+\lambda)};$$

also that

$$\frac{x}{(a+\alpha)^2}+\frac{y}{(b+\alpha)^2}+\frac{z}{(c+\alpha)^2}=\frac{(\gamma-\alpha)(\alpha-\beta)}{(a+\alpha)(b+\alpha)(c+\alpha)}.$$

CHAPTER IV.

ON THE CONIC REPRESENTED BY THE GENERAL EQUATION OF THE SECOND DEGREE.

1. We may now proceed to the discussion of the general equation of the second degree, which we shall express under the form,

$$u\alpha^2 + v\beta^2 + w\gamma^2 + 2u'\beta\gamma + 2v'\gamma\alpha + 2w'\alpha\beta = 0.$$

This we may write, for shortness, $\phi(\alpha, \beta, \gamma) = 0$.

This equation, as we have shewn (Art. 1, Chap. II.), represents a conic section.

2. *To find the point in which a straight line, drawn in a given direction through a given point of the conic, meets the conic again.*

Let f, g, h be the co-ordinates of the given point, α, β, γ those of any other point whatever. Then, for all points of the straight line joining these two, the quantities

$$\alpha - f, \quad \beta - g, \quad \gamma - h,$$

will bear constant ratios to one another. Let these ratios be denoted by $p : q : r$, so that we have

$$\frac{\alpha - f}{p} = \frac{\beta - g}{q} = \frac{\gamma - h}{r} = s, \text{ suppose.}$$

EQUATION OF THE TANGENT.

To find where the line again meets the conic, we must substitute in the equation of the conic

$$f+ps \text{ for } \alpha, \quad g+qs \text{ for } \beta, \quad h+rs \text{ for } \gamma.$$

We thus get, arranging the result according to ascending powers of s,

$$\phi(f, g, h) + 2\{(up+w'q+v'r)f + (w'p+vq+u'r)g$$
$$+ (v'p+u'q+wr)h\}\, s + \phi(p, q, r)\, s^2 = 0.$$

The two roots of this equation, considered as a quadratic in s, determine the two points where the line meets the conic.

Now, since (f, g, h) is, by supposition, a point on the conic, it follows that $\phi(f, g, h)$ must be itself $= 0$. Hence, one of the two values of s, given by the above equation, will $= 0$, as ought to be the case, this value corresponding to the point f, g, h itself. The value of s, corresponding to the other point of intersection, will then be

$$-2\frac{(up+w'q+v'r)f + (w'p+vq+u'r)g + (v'p+u'q+wr)h}{\phi(p, q, r)}.$$

Hence, the values of α, β, γ, may be determined.

To this value of s, we shall hereafter have occasion to refer.

3. *To find the equation of the tangent at a given point.*

If the two points in which a straight line meets the conic be indefinitely close together, the value of s, investigated in Art. 2, must be $= 0$. This gives

$$(up+w'q+v'r)f + (w'p+vq+u'r)g + (v'p+u'q+wr)h = 0,$$

or,

$$(uf+w'g+v'h)p + (w'f+vg+u'h)q + (v'f+u'g+wh)r = 0.$$

Hence, since, for every point on the line required,

$$\frac{\alpha-f}{p} = \frac{\beta-g}{q} = \frac{\gamma-h}{r},$$

we get

$$(uf + w'g + v'h)\alpha + (w'f + vg + u'h)\beta + (v'f + u'g + wh)\gamma$$
$$= uf^2 + vg^2 + wh^2 + 2u'gh + 2v'hf + 2w'fg$$
$$= 0, \text{ since } (f, g, h) \text{ is a point on the conic.}$$

The tangent, therefore, at (f, g, h) is represented by the equation

$$(uf + w'g + v'h)\alpha + (w'f + vg + u'h)\beta + (v'f + u'g + wh)\gamma = 0.$$

OBS. Those who are acquainted with the Differential Calculus will remark that this equation may be written thus,

$$\frac{d\phi}{df}\alpha + \frac{d\phi}{dg}\beta + \frac{d\phi}{dh}\gamma = 0.$$

4. *To find the condition that a given straight line may touch the conic.*

Let the equation of the given straight line be

$$l\alpha + m\beta + n\gamma = 0.$$

Let (f, g, h) be the co-ordinates of its point of contact; then, comparing this with the equation of the tangent just investigated, we see that we must have

$$\frac{uf + w'g + v'h}{l} = \frac{w'f + vg + u'h}{m} = \frac{v'f + u'g + wh}{n}.$$

Representing each of these equivalent quantities by $-k$, we shall have

$$uf + w'g + v'h + lk = 0 \quad \ldots\ldots\ldots\ldots\ldots (1),$$
$$w'f + vg + u'h + mk = 0 \quad \ldots\ldots\ldots\ldots\ldots (2),$$
$$v'f + u'g + wh + nk = 0 \quad \ldots\ldots\ldots\ldots\ldots (3).$$

Also, since (f, g, h) is a point on the given line,

$$lf + mg + nh = 0 \quad \ldots\ldots\ldots\ldots (4).$$

CONDITION OF TANGENCY.

Eliminating f, g, h, k, between (1), (2), (3), (4), we obtain

$$\begin{vmatrix} u, & w', & v', & l \\ w', & v, & u', & m \\ v', & u', & w, & n \\ l, & m, & n, & 0 \end{vmatrix} = 0, \text{ or } \begin{vmatrix} \dfrac{d^2\phi}{d\alpha^2}, & \dfrac{d^2\phi}{d\alpha d\beta}, & \dfrac{d^2\phi}{d\alpha d\gamma}, & l \\ \dfrac{d^2\phi}{d\beta d\alpha}, & \dfrac{d^2\phi}{d\beta^2}, & \dfrac{d^2\phi}{d\beta d\gamma}, & m \\ \dfrac{d^2\phi}{d\gamma d\alpha}, & \dfrac{d^2\phi}{d\gamma d\beta}, & \dfrac{d^2\phi}{d\gamma^2}, & n \\ l, & m, & n, & 0 \end{vmatrix} = 0$$

as the necessary condition that the line (l, m, n) should touch the conic $\phi(f, g, h) = 0$. Expanding the determinant, this may be written

$$(vw - u'^2) l^2 + (wu - v'^2) m^2 + (uv - w'^2) n^2 + 2 (v'w' - uu') mn$$
$$+ 2 (w'u' - vv') nl + 2 (u'v' - ww') lm = 0.$$

5. The coefficients of l^2, m^2, n^2, $2mn$, $2nl$, $2lm$, in the above equation, will be observed to be the several minors of the determinant

$$\begin{vmatrix} u, & w', & v' \\ w', & v, & u' \\ v', & u', & w \end{vmatrix}$$

They will frequently present themselves in subsequent investigations, and it will be convenient, therefore, to denote each by a single letter. We shall adopt the following notation:

$$vw - u'^2 = U, \quad wu - v'^2 = V, \quad uv - w'^2 = W,$$
$$v'w' - uu' = U', \quad w'u' - vv' = V', \quad u'v' - ww' = W'.$$

The condition of tangency investigated in Art. 4 may then be written,

$$Ul^2 + Vm^2 + Wn^2 + 2U'mn + 2V'nl + 2W'lm = 0,$$

the same condition, it will be observed, as that which must hold, in order that the point (l, m, n) may lie on the conic

$$U\alpha^2 + V\beta^2 + W\gamma^2 + 2U'\beta\gamma + 2V'\gamma\alpha + 2W'\alpha\beta = 0.$$

6. *To find the condition that the conic may be a parabola.*

Since every parabola touches the line at infinity, the condition required will be obtained by writing a, b, c respectively in place of l, m, n, in the condition of tangency. This gives, as the necessary and sufficient relation among the coefficients,

$$\begin{vmatrix} u, & w', & v', & a \\ w', & v, & u', & b \\ v', & u', & w, & c \\ a, & b, & c, & 0 \end{vmatrix} = 0$$

or $Ua^2 + Vb^2 + Wc^2 + 2U'bc + 2V'ca + 2W'ab = 0.$

7. *To find the condition that the conic may break up into two straight lines, real or imaginary.*

For this purpose it is necessary and sufficient that the expression $\phi(\alpha, \beta, \gamma)$ should break up into two factors. The condition for this has been shewn in Art. 11, Chap. III. to be

$$\begin{vmatrix} u, & w', & v' \\ w', & v, & u' \\ v', & u', & w \end{vmatrix} = 0$$

or $uvw + 2u'v'w' - uu'^2 - vv'^2 - ww'^2 = 0.$

8. *To find the equation of the polar of a given point.*

If through a given point any straight line be drawn cutting a conic in two points, and at each point of section a tangent be drawn to the curve, the locus of the intersection of these tangents is the polar of the given point. We proceed to find the equation of the polar of (f, g, h).

$f_1, g_1, h_1, f_2, g_2, h_2$ be the co-ordinates of the points in which any straight line drawn through (f, g, h) meets the conic. Then, since (f, g, h), (f_1, g_1, h_1), (f_2, g_2, h_2) lie in the same straight line, we have

$$f(g_1 h_2 - g_2 h_1) + g(h_1 f_2 - h_2 f_1) + h(f_1 g_2 - f_2 g_1) = 0 \ldots \ldots (1)$$

(see Art. 12, Chap. I.). Again, the equations of the tangents at (f_1, g_1, h_1), (f_2, g_2, h_2) respectively, are

$$f_1(u\alpha + w'\beta + v'\gamma) + g_1(w'\alpha + v\beta + u'\gamma) + h_1(v'\alpha + u'\beta + w\gamma) = 0,$$
$$f_2(u\alpha + w'\beta + v'\gamma) + g_2(w'\alpha + v\beta + u'\gamma) + h_2(v'\alpha + u'\beta + w\gamma) = 0.$$

Where these intersect, we have

$$\frac{u\alpha + w'\beta + v'\gamma}{g_1 h_2 - g_2 h_1} = \frac{w'\alpha + v\beta + u'\gamma}{h_1 f_2 - h_2 f_1} = \frac{v'\alpha + u'\beta + w\gamma}{f_2 g_2 - f_2 g_1} \ldots \ldots (2).$$

Combining this with equation (1), we get

$$f(u\alpha + w'\beta + v'\gamma) + g(w'\alpha + v\beta + u'\gamma) + h(v'\alpha + u'\beta + wg) = 0$$

or $(uf + w'g + v'h)\alpha + (w'f + vg + u'h)\beta + (v'f + u'g + wh)\gamma = 0$,

as a relation which holds at the intersection of the tangents; and which, since it is independent of the values of f_1, g_1, h_1; f_2, g_2, h_2; must be the equation of the locus of the point of intersection of the tangents drawn at the extremities of *any* chord passing through (f, g, h), that is, it is the equation of the polar of (f, g, h).

It may also be written,

$$\frac{d\phi}{df}\alpha + \frac{d\phi}{dg}\beta + \frac{d\phi}{dh}\gamma = 0.$$

It will be remarked that this equation is identical in form with that already investigated for the tangent at a point (f, g, h) of the curve. In fact, when the point (f, g, h) is on the curve, the polar and the tangent become identical.

9. *To find the co-ordinates of the pole of a given straight line.*

Let the equation of the given straight line be

$$l\alpha + m\beta + n\gamma = 0.$$

If (f, g, h) be the co-ordinates of its pole, we must have, applying the equation just investigated for the polar of (f, g, h),

$$\frac{uf + w'g + v'h}{l} = \frac{w'f + vg + u'h}{m} = \frac{v'f + u'g + wh}{n}.$$

Putting each member of these equations $= -k$, we get

$$uf + w'g + v'h + \ lk = 0,$$
$$w'f + \ vg + u'h + mk = 0,$$
$$v'f + u'g + wh + nk = 0,$$

whence

$$\frac{f}{\begin{vmatrix} w', & v', & l \\ v, & u', & m \\ u', & w, & n \end{vmatrix}} = \frac{g}{\begin{vmatrix} u', & w', & m \\ w, & v', & n \\ v', & u, & l \end{vmatrix}} = \frac{h}{\begin{vmatrix} v', & u', & n \\ u, & w', & l \\ w', & v, & m \end{vmatrix}}$$

These equations, together with

$$af + bg + ch = 2\Delta,$$

determine the co-ordinates of the pole. They may also be written

$$\frac{f}{Ul + W'm + V'n} = \frac{g}{W'l + Vm + U'n} = \frac{h}{V'l + U'm + Wn}.$$

10. *To find the equation of the pair of tangents drawn to the conic from a given external point.*

Consider the equation

$$\phi(\alpha, \beta, \gamma) + k\{(uf + w'g + v'h)\alpha + (w'f + vg + u'h)\beta + (v'f + u'g + wh)\gamma\}^2 = 0,$$

where k is an arbitrary constant.

This, being of the second degree in α, β, γ, represents a conic; and meets the conic $\phi(\alpha, \beta, \gamma) = 0$ in the two points in which that conic meets the line

$$(uf + w'g + v'h)\alpha + (w'f + vg + u'h)\beta + (v'f + u'g + wh)\gamma = 0,$$

POLE AND POLAR. 81

and in these points only. Hence since two conics in general intersect in four points, it follows that in this case the four points of intersection coincide two and two, that is, the conics touch one another at the two points where they meet the above-mentioned line, or have *double contact* with each other.

The arbitrary constant k may be determined by making the conic pass through any assigned point. Suppose now that the conic is required to pass through the point (f, g, h), of which the line of contact is the polar. This gives, for the determination of k, the condition

$$\phi(f, g, h) + k\{(uf + w'g + v'h)f + (w'f + vg + u'h)g \\ + (v'f + u'g + wh)h\}^2 = 0,$$

whence $$k = -\frac{1}{\phi(f, g, h)}.$$

Hence the equation

$$\phi(f, g, h)\phi(\alpha, \beta, \gamma) - \{(uf + w'g + v'h)\alpha \\ + (w'f + vg + u'h)\beta + (v'f + u'g + wh)\gamma\}^2 = 0,$$

represents the curve of the second degree, passing through the point (f, g, h) and touching the conic $\phi(\alpha, \beta, \gamma) = 0$, at the points where the polar of this point intersects it. But this curve must evidently be coincident with the two tangents drawn from that point to the given conic $\phi(\alpha, \beta, \gamma) = 0$.

This equation may be put under another form, for the coefficient of α^2 will be found, by actual expansion, to be

$$u(uf^2 + vg^2 + wh^2 + 2u'gh + 2v'hf + 2w'fg) \\ - (u^2f^2 + w'^2g^2 + v'^2h^2 + 2v'w'gh + 2uv'hf + 2uw'fg) \\ = (uv - w'^2)g^2 + (wu - v'^2)h^2 + 2(uu' - v'w')gh \\ = Wg^2 + Vh^2 - 2U'gh.$$

That of $2\beta\gamma$ is

$$u'(uf^2 + vg^2 + wh^2 + 2u'gh + 2v'hf + 2w'fg) \\ - (w'f + vg + u'h)(v'f + u'g + wh) \\ = (uu' - v'w')f^2 + (u'^2 - vw)gh + (u'v' - ww')hf + (w'u' - vv')fg \\ = -U'f^2 - Ugh + W'hf + V'fg.$$

F. 6

Similar expressions holding for the coefficients of β^2, γ^2, $2\gamma\alpha$, $2\alpha\beta$, we obtain the equation of the two tangents under the form

$$(Wg^2+Vh^2-2U'gh)\alpha^2+(Uh^2+Wf^2-2V'hf)\beta^2+(Vf^2+Ug^2-2W'fg)\gamma^2$$
$$-2(U'f^2+Ugh-W'hf-V'fg)\beta\gamma$$
$$-2(V'g^2+Vhf-U'fg-W'gh)\gamma\alpha$$
$$-2(W'h^2+Wfg-V'gh-U'hf)\alpha\beta=0.$$

If the point (f, g, h) be *within* the conic, these two tangents will be imaginary.

11. *To find the co-ordinates of the centre.*

Since the two tangents drawn at the extremities of any chord passing through the centre, are parallel to each other, it follows that the polar of the centre is at an infinite distance, and may therefore be represented by the equation

$$a\alpha + b\beta + c\gamma = 0.$$

Hence, if $\bar{\alpha}$, $\bar{\beta}$, $\bar{\gamma}$, be the co-ordinates of the centre, we obtain, by an investigation similar to that of Art. 9,

$$\left. \begin{array}{r} u\bar{\alpha} + w'\bar{\beta} + v'\bar{\gamma} + ak = 0, \\ w'\bar{\alpha} + v\bar{\beta} + u'\bar{\gamma} + bk = 0, \\ v'\bar{\alpha} + u'\bar{\beta} + w\bar{\gamma} + ck = 0, \end{array} \right\} \ldots\ldots(A).$$

Hence,

$$\frac{\bar{\alpha}}{\begin{vmatrix} w', v', a \\ v, u', b \\ u', w, c \end{vmatrix}} = \frac{\bar{\beta}}{\begin{vmatrix} u', w', b \\ w, v', c \\ v', u, a \end{vmatrix}} = \frac{\bar{\gamma}}{\begin{vmatrix} v', u', c \\ u, w', a \\ w', v, b \end{vmatrix}} = -\frac{k}{\begin{vmatrix} u, w', v' \\ w', v, u' \\ v', u', w \end{vmatrix}},$$

or

$$\frac{\bar{\alpha}}{Ua+W'b+V'c} = \frac{\bar{\beta}}{W'a+Vb+U'c} = \frac{\bar{\gamma}}{V'a+U'b+Wc}$$
$$= -\frac{k}{uvw+2u'v'w'-uu'^2-vv'^2-ww'^2}.$$

These equations determine the centre.

ASYMPTOTES.

12. *To find the equation of the asymptotes.*

Writing $\bar{\alpha}, \bar{\beta}, \bar{\gamma}$, for f, g, h, in the investigation of Art. 10, and paying regard to equations (A) of Art. 11, the asymptotes will be found to be represented by the equation

$$\phi(\bar{\alpha}, \bar{\beta}, \bar{\gamma}) \phi(\alpha, \beta, \gamma) - \{(a\alpha + b\beta + c\gamma)^2\} k^2 = 0,$$

or $\quad\phi(\bar{\alpha}, \bar{\beta}, \bar{\gamma}) \phi(\alpha, \beta, \gamma) - (2\Delta)^2 k^2 = 0.$

But, multiplying equations (A) in order by $\bar{\alpha}, \bar{\beta}, \bar{\gamma}$, and adding, we get

$$\phi(\bar{\alpha}, \bar{\beta}, \bar{\gamma}) + 2\Delta \cdot k = 0.$$

Hence the asymptotes may be represented by the equation

$$\phi(\alpha, \beta, \gamma) - \phi(\bar{\alpha}, \bar{\beta}, \bar{\gamma}) = 0,$$

or $\quad\phi(\alpha, \beta, \gamma) + 2\Delta \cdot k = 0,$

which may be put under the homogeneous form

$$(a\bar{\alpha} + b\bar{\beta} + c\bar{\gamma}) \phi(\alpha, \beta, \gamma) + k(a\alpha + b\beta + c\gamma)^2 = 0.$$

But, by the final result of Art. 11, it may be seen that

$$\frac{a\bar{\alpha} + b\bar{\beta} + c\bar{\gamma}}{k} = -\frac{Ua^2 + Vb^2 + Wc^2 + 2U'bc + 2V'ca + 2W'ab}{uvw + 2u'v'w' - uu'^2 - vv'^2 - ww'^2},$$

whence the equation of the asymptotes becomes

$$(Ua^2 + Vb^2 + Wc^2 + 2U'bc + 2V'ca + 2W'ab) \phi(\alpha, \beta, \gamma)$$
$$- (uvw + 2u'v'w' - uu'^2 - vv'^2 - ww'^2)(a\alpha + b\beta + c\gamma)^2 = 0.$$

This may also be written under the form

$$\begin{vmatrix} u, & w', & v', & a \\ w', & v, & u', & b \\ v', & u', & w, & c \\ a, & b, & c, & 0 \end{vmatrix} \phi(\alpha, \beta, \gamma) + \begin{vmatrix} u, & w', & v' \\ w', & v, & u' \\ v', & u', & w \end{vmatrix} (a\alpha + b\beta + c\gamma)^2 = 0.$$

Cor. It appears, from the preceding investigation, that if $\bar{\alpha}, \bar{\beta}, \bar{\gamma}$ be the co-ordinates of the centre of the conic represented by the equation

$$\phi(\alpha, \beta, \gamma) = u\alpha^2 + v\beta^2 + w\gamma^2 + 2u'\beta\gamma + 2v'\gamma\alpha + 2w'\alpha\beta = 0,$$

then $\phi(\bar{\alpha}, \bar{\beta}, \bar{\gamma}) = - \dfrac{\begin{vmatrix} u, & w', & v' \\ w', & v, & u' \\ v', & u', & w \end{vmatrix} 4\Delta^2}{\begin{vmatrix} u, & w', & v', & a \\ w', & v, & u', & b \\ v', & u', & w, & c \\ a, & b, & c, & 0 \end{vmatrix}}.$

13. *To find the condition that the conic may be a rectangular hyperbola.*

If the equations of the asymptotes be

$$l\alpha + m\beta + n\gamma = 0,$$
$$l'\alpha + m'\beta + n'\gamma = 0,$$

the condition of their perpendicularity is

$$ll' + mm' + nn' - (mn' + m'n) \cos A - (nl' + n'l) \cos B$$
$$- (lm' + l'm) \cos C = 0.$$

Writing, for shortness,

$$Ua^2 + Vb^2 + Wc^2 + 2U'bc + 2V'ca + 2W'ab = D,$$
$$uvw + 2u'v'w' - uu'^2 - vv'^2 - ww'^2 = K,$$

we see, by reference to Art. 12, that

$$\frac{ll'}{Du - Ka^2} = \frac{mm'}{Dv - Kb^2} = \frac{nn'}{Dw - Kc^2}$$

$$= \frac{\tfrac{1}{2}(mm' + m'n)}{Du' - Kbc} = \frac{\tfrac{1}{2}(nl' + n'l)}{Dv' - Kca} = \frac{\tfrac{1}{2}(lm' + l'm)}{Dw' - Kab}.$$

Hence the required condition is

$$D(u + v + w - 2u' \cos A - 2v' \cos B - 2w' \cos C)$$
$$- K(a^2 + b^2 + c^2 - 2bc \cos A - 2ca \cos B - 2ab \cos C) = 0.$$

Now $a^2 + b^2 + c^2 - 2bc \cos A - 2ca \cos B - 2ab \cos C = 0$ identically, hence the required condition becomes

$$u + v + w - 2u' \cos A - 2v' \cos B - 2w' \cos C = 0.$$

COR. It hence appears, that the condition that the conic

$$u'\beta\gamma + v'\gamma\alpha + w'\alpha\beta = 0,$$

described about the triangle of reference, may be a rectangular hyperbola, is

$$u' \cos A + v' \cos B + w' \cos C = 0;$$

that is, the conic must pass through the point determined by the equations

$$\alpha \cos A = \beta \cos B = \gamma \cos C.$$

This point (see Art. 5, Chap. I.) is the point of intersection of the perpendiculars let fall from each angular point of the triangle on the opposite side. Hence we obtain the following elegant geometrical proposition, that *every rectangular hyperbola described about a given triangle passes through the point of intersection of the perpendiculars let fall from each angular point of the triangle on the opposite side.*

Again, if u', v', w' be all $= 0$, the condition is

$$u + v + w = 0,$$

proving that, if the equation

$$u\alpha^2 + v\beta^2 + w\gamma^2 = 0$$

represent a rectangular hyperbola, the curve will pass through the four points for which

$$\alpha = \beta = \gamma, \ -\alpha = \beta = \gamma, \ \alpha = -\beta = \gamma, \ \alpha = \beta = -\gamma.$$

In other words, *if a rectangular hyperbola be so described that each angular point of a given triangle is the pole, with respect to it, of the opposite side, it will pass through the centres of the four circles which touch the three sides of the triangle.*

14. *To investigate the conditions that the general equation of the second degree shall represent a circle.*

The property of the circle, which we shall assume as the basis of our investigation, is the following: that if, through any point, chords be drawn cutting a circle, the rectangle, contained by their segments, is invariable.

Suppose then, that the curve, represented by the equation
$$u\alpha^2 + v\beta^2 + w\gamma^2 + 2u'\beta\gamma + 2v'\gamma\alpha + 2w'\alpha\beta = 0,$$
cut BC in b_1, c_1, CA in c_2, a_2, AB in a_3, b_3,

Fig. 17.

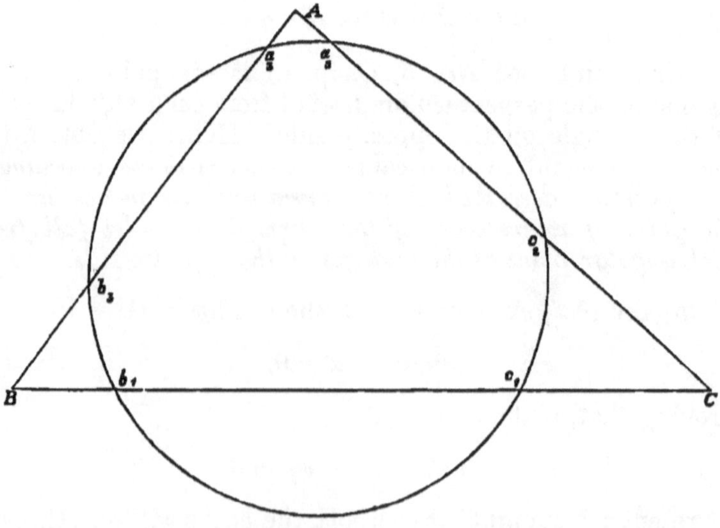

then, if this curve be a circle,
$$Ac_2 \cdot Aa_2 = Aa_3 \cdot Ab_3, \ Ba_3 \cdot Bb_3 = Bb_1 \cdot Bc_1, \ Cb_1 \cdot Cc_1 = Cc_2 \cdot Ca_2.$$

CONDITIONS FOR A CIRCLE.

Let h, h' be the respective distances of c_2, a_2 from AB; g, g' those of a_3, b_3 from AC: then multiplying the first of the above three equations by $\sin^2 A$, we get

$$hh' = gg'.$$

Now h, h' are the two values of γ obtained by putting $\beta = 0$ in the equation of the conic section, bearing in mind that, when $\beta = 0$,

$$a\alpha + c\gamma = 2\Delta.$$

This gives, for the determination of γ, the equation

$$u(c\gamma - 2\Delta)^2 + wa^2\gamma^2 + 2av'\gamma(2\Delta - c\gamma) = 0;$$

whence, by the theory of equations,

$$hh' = \frac{u \cdot 4\Delta^2}{uc^2 + wa^2 - 2v'ca}.$$

Similarly,
$$gg' = \frac{u \cdot 4\Delta^2}{va^2 + ub^2 - 2w'ab}.$$

Hence, since $Ac_2 \cdot Aa_2 = Aa_3 \cdot Ab_3$, we obtain

$$uc^2 + wa^2 - 2v'ca = va^2 + ub^2 - 2w'ab.$$

Similarly, from the condition

$$Ba_3 \cdot Bb_3 = Bb_1 \cdot Bc_1$$

we find
$$va^2 + ub^2 - 2w'ab = wb^2 + vc^2 - 2u'bc.$$

The condition $Cb_1 \cdot Cc_1 = Cc_2 \cdot Ca_2$

gives
$$wb^2 + vc^2 - 2u'bc = uc^2 + wa^2 - 2v'ca,$$

which also follows from the preceding two equations. Hence the equations

$$wb^2 + vc^2 - 2u'bc = uc^2 + wa^2 - 2v'ca = va^2 + ub^2 - 2w'ab$$

are necessary conditions that the given equation should represent a circle; and, since they are two in number, they are sufficient.

88 TRILINEAR CO-ORDINATES.

15. *To determine the intersection of a circle with the line at infinity.*

Since, at every point in the line at infinity,
$$a\alpha + b\beta + c\gamma = 0,$$
we shall have
$$\alpha^2 = -\frac{c\gamma\alpha + b\alpha\beta}{a},$$

$$\beta^2 = -\frac{a\alpha\beta + c\beta\gamma}{b},$$

$$\gamma^2 = -\frac{b\beta\gamma + a\gamma\alpha}{c}.$$

Substituting these values in the equation
$$u\alpha^2 + v\beta^2 + w\gamma^2 + 2u'\beta\gamma + 2v'\gamma\alpha + 2w'\alpha\beta = 0,$$
we get
$$\left(2u' - \frac{vc}{b} - \frac{wb}{c}\right)\beta\gamma + \left(2v' - \frac{wa}{c} - \frac{uc}{a}\right)\gamma\alpha$$
$$+ \left(2w' - \frac{ub}{a} - \frac{va}{b}\right)\alpha\beta = 0;$$
or, multiplying by abc,
$$(2u'bc - vc^2 - wb^2)\alpha\beta\gamma + (2v'ca - wa^2 - uc^2)b\gamma\alpha$$
$$+ (2w'ab - ub^2 - va^2)c\alpha\beta = 0,$$
which, if the conic be a circle, reduces to
$$\alpha\beta\gamma + b\gamma\alpha + c\alpha\beta = 0,$$
shewing that every circle intersects the line at infinity in the same two points as the circle described about the triangle of reference; that is, *all circles intersect the line at infinity in the same two points.* These points are, of course, imaginary.

From this it follows that every circle may be represented in either of the following forms,
$$a\beta\gamma + b\gamma\alpha + c\alpha\beta + (l\alpha + m\beta + n\gamma)(a\alpha + b\beta + c\gamma) = 0,$$
$$\sin 2A \cdot \alpha^2 + \sin 2B \cdot \beta^2 + \sin 2C \cdot \gamma^2 + (\lambda\alpha + \mu\beta + \nu\gamma)(a\alpha + b\beta + c\gamma) = 0.$$

16. It may be shewn, by a geometrical investigation similar to that in Art. 14, that if ρ_1, ρ_2, ρ_3 be the semi-diameters of the conic respectively parallel to the sides of the triangle of reference,

$$\rho_1^2 (wb^2 + vc^2 - 2u'bc) = \rho_2^2 (uc^2 + wa^2 - 2v'ca)$$
$$= \rho_3^2 (va^2 + ub^2 - 2w'ab).$$

Hence, if two conics be similar and similarly situated, the values of the ratios denoted by

$$wb^2 + vc^2 - 2u'bc : uc^2 + wa^2 - 2v'ca : va^2 + ub^2 - 2w'ab$$

must be the same for both.

Hence, also, by reasoning similar to that employed in Art. 15, it follows that *all conics, similar and similarly situated to each other, intersect in the same two points in the line at infinity.*

These points will be real, coincident, or imaginary, according as the conics are hyperbolas, parabolas, or ellipses.

If the conics, in addition to being similar and similarly situated, are also concentric, they will touch one another at the two points where they meet the line at infinity.

17. *To find the radical axis of two similar and similarly situated conics.*

By multiplying the equation of one of two given conics by an arbitrary constant, and adding it to the equation of the other given conics, we obtain the general equation of the system of conics passing through their four points of intersection. By suitably determining the arbitrary constant, we may make this equation represent any one of the three pairs of straight lines passing through these four points. In the case, therefore, in which the two conics are similar and similarly situated, it must be possible so to determine the constant that the left-hand member of the equation may break up into two factors, one which equated to zero represents the line at infinity, and the other the radical axis. Hence, if

$$u\alpha^2 + v\beta^2 + w\gamma^2 + 2u'\beta\gamma + 2v'\gamma\alpha + 2w'\alpha\beta = 0,$$

$$p\alpha^2 + q\beta^2 + r\gamma^2 + 2p'\beta\gamma + 2q'\gamma\alpha + 2r'\alpha\beta = 0,$$

be the equations of two similar and similarly situated conics, it must be possible to determine the arbitrary multiplier k, so that

$$(u + kp)\, \alpha^2 + (v + kq)\, \beta^2 + (w + kr)\, \gamma^2$$
$$+ 2\,(u' + kp')\, \beta\gamma + 2\,(v' + kq')\, \gamma\alpha + 2\,(w' + kr')\, \alpha\beta$$
$$= (a\alpha + b\beta + c\gamma) \left(\frac{u + kp}{a} \alpha + \frac{v + kq}{b} \beta + \frac{w + kr}{c} \gamma \right)$$

identically.

This gives, equating the coefficients of $\beta\gamma$, $\gamma\alpha$, $\alpha\beta$,

$$2\,(u' + kp') = (v + kq)\, \frac{c}{b} + (w + kr)\, \frac{b}{c},$$

$$2\,(v' + kq') = (w + kr)\, \frac{a}{c} + (u + kp)\, \frac{c}{a},$$

$$2\,(w' + kr') = (u + kp)\, \frac{b}{a} + (v + kq)\, \frac{a}{b};$$

$$\therefore k = -\frac{wb^2 + vc^2 - 2u'bc}{rb^2 + qc^2 - 2p'bc} = -\frac{uc^2 + wa^2 - 2v'ca}{pc^2 + ra^2 - 2q'ca}$$

$$= -\frac{va^2 + ub^2 - 2w'ab}{qa^2 + pb^2 - 2r'ab}.$$

(The identity of these three values of k is ensured by the condition of similarity already investigated.)

k may also be written

$$-\frac{\dfrac{u}{a^2} + \dfrac{v}{b^2} + \dfrac{w}{c^2} - \dfrac{u'}{bc} - \dfrac{v'}{ca} - \dfrac{w'}{ab}}{\dfrac{p}{a^2} + \dfrac{q}{b^2} + \dfrac{r}{c^2} - \dfrac{p'}{bc} - \dfrac{q'}{ca} - \dfrac{r'}{ab}}.$$

Hence, the equation of the radical axis becomes

$$\frac{\dfrac{u\alpha}{a}+\dfrac{v\beta}{b}+\dfrac{w\gamma}{c}}{\dfrac{u}{a^2}+\dfrac{v}{b^2}+\dfrac{w}{c^2}-\dfrac{u'}{bc}-\dfrac{v'}{ca}-\dfrac{w'}{ab}} = \frac{\dfrac{p\alpha}{a}+\dfrac{q\beta}{b}+\dfrac{r\gamma}{c}}{\dfrac{p}{a^2}+\dfrac{q}{b^2}+\dfrac{r}{c^2}-\dfrac{p'}{bc}-\dfrac{q'}{ca}-\dfrac{r'}{ab}}.$$

18. As an example of the application of this formula we may take the following theorem. *The nine-point circle of a triangle (that is, the circle which passes through the middle points of its sides) touches each of the four circles which touch the three sides of the triangle.*

Suppose that $\quad \lambda\alpha + \mu\beta + \nu\gamma = 0$

is the equation of the radical axis of the inscribed and nine-point circles. The equation of the nine-point circle will then be (see Chap. II. Art. 10),

$$a^2(s-a)^2\alpha^2 + b^2(s-b)^2\beta^2 + c^2(s-c)^2\gamma^2$$
$$- 2bc(s-b)(s-c)\beta\gamma - 2ca(s-c)(s-a)\gamma\alpha$$
$$- 2ab(s-a)(s-b)\alpha\beta + (\lambda\alpha + \mu\beta + \nu\gamma)(a\alpha + b\beta + c\gamma) = 0.$$

If this represent the nine-point circle, it must be satisfied when $\alpha = 0$ and $b\beta = c\gamma$. Hence

$$(s-b)^2 + (s-c)^2 - 2(s-b)(s-c) + 2\left(\frac{\mu}{b}+\frac{\nu}{c}\right) = 0,$$

$$\text{or} \quad \frac{\mu}{b}+\frac{\nu}{c} = \frac{(b-c)^2}{2}.$$

Similarly $\quad \dfrac{\nu}{c}+\dfrac{\lambda}{a} = \dfrac{(c-a)^2}{2},$

$$\frac{\lambda}{a}+\frac{\mu}{b} = \frac{(a-b)^2}{2};$$

$$\therefore \quad \frac{2\lambda}{a} = \frac{(c-a)^2 + (a-b)^2 - (b-c)^2}{2}$$

$$= (a-b)(a-c);$$

$\therefore \quad \lambda = \tfrac{1}{2}a(a-b)(a-c).$

Similarly $\quad \mu = \tfrac{1}{2}b(b-c)(b-a),$

$\quad \nu = \tfrac{1}{2}c(c-a)(c-b).$

This gives, for the equation of the radical axis,

$$\frac{a\alpha}{b-c} + \frac{b\beta}{c-a} + \frac{c\gamma}{a-b} = 0.$$

Now, to ascertain whether this touches the inscribed circle, we have, applying the condition of Chap. II. Art. 9, to investigate the value of

$$\frac{b-c}{a}\cos^2\frac{A}{2} + \frac{c-a}{b}\cos^2\frac{B}{2} + \frac{a-b}{c}\cos^2\frac{C}{2},$$

or $\quad\dfrac{s}{abc}\{(b-c)(s-a) + (c-a)(s-b) + (a-b)(s-c)\}$,

which is 0. Hence, the radical axis touches the inscribed circle, and therefore the inscribed and nine-point circles touch one another. Similarly, it may be proved that the nine-point circle touches each of the escribed circles.

19. The equation of the nine-point circle may be deduced by substituting the above values of λ, μ, ν, or (perhaps more neatly) by expressing the fact that the curve

$$u\alpha^2 + v\beta^2 + w\gamma^2 + 2u'\beta\gamma + 2v'\gamma\alpha + 2w'\alpha\beta = 0$$

passes through the middle points of the sides of the triangle, and combining the equations thus obtained with those investigated in Art. 14. The former gives

$$vc^2 + wb^2 + 2u'bc = 0,$$
$$wa^2 + uc^2 + 2v'ca = 0,$$
$$ub^2 + va^2 + 2w'ab = 0.$$

Hence, by Art. 14, $u'bc = v'ca = w'ab$.

Supposing $u' = -a$, we get

$$\frac{v}{b^2} + \frac{w}{c^2} = \frac{2a}{bc},$$

with two similar equations, whence

$$\frac{u}{a^2} = \frac{b^2 + c^2 - a^2}{abc} = \frac{2\cos A}{a};$$

$$\therefore\ u = 2a\cos A.$$

INTERSECTION OF RECTANGULAR TANGENTS.

Hence, the nine-point circle is represented by the equation
$$a \cos A \cdot \alpha^2 + b \cos B \cdot \beta^2 + c \cos C \cdot \gamma^2 - a\beta\gamma - b\gamma\alpha - c\alpha\beta = 0.$$

COR. It hence appears that the nine-point circle passes through the points of intersection of the circumscribed and self-conjugate circles, or has a common radical axis with them.

20. We have investigated, in Art. 10, the equation of the pair of tangents drawn to the conic from a given point (f, g, h). If these two tangents be at right angles to one another, they may be regarded as the limiting form of a rectangular hyperbola, and must therefore satisfy the equation investigated in Art. 13. This, therefore, gives as the locus of the intersection of two tangents at right angles to one another

$$Wg^2 + Vh^2 - 2U'gh + Uh^2 + Wf^2 - 2V'hf + Vf^2 + Ug^2 - 2W'fg$$
$$+ 2(Uf^2 + Ugh - V'fg - W'hf)\cos A$$
$$+ 2(V'g^2 + Vhf - W'gh - U'fg)\cos B$$
$$+ 2(W'h^2 + Wfg - U'hf - V'gh)\cos C = 0.$$

This may be shewn (see Art. 15) to represent a circle, as we know ought to be the case.

This equation may also be expressed in the following form:

$$(af + bg + ch)\left(\frac{V + W + 2U'\cos A}{a}f + \frac{W + U + 2V'\cos B}{b}g + \frac{U + V + 2W'\cos C}{c}h\right)$$
$$- \left(\frac{Ua^2 + Vb^2 + Wc^2 + 2U'bc + 2V'ca + 2W'ab}{abc}\right)(agh + bhf + cfg) = 0.$$

If the conic be a parabola, then (see Art. 6) this breaks up into two factors, one of which is the line at infinity; and the other must represent the directrix, since that is the locus of the point of intersection of two tangents to a parabola at right angles to one another.

The appearance of the line at infinity as a factor in the result in this case may be explained as follows: Every parabola touches the line at infinity, and this line also satisfies the algebraical condition of being perpendicular to any line whatever, since, whatever l, m, n may be,

$al+bm+cn-(bn+cm)\cos A-(cl+an)\cos B-(am+bl)\cos C=0$,
identically.

It therefore will form a part of the locus of the intersection of two tangents at right angles to one another, the two tangents being the line at infinity itself, and any other tangent whatever.

The directrix of the parabola is therefore represented by the equation

$$\frac{V+W+2U'\cos A}{a}\alpha+\frac{W+U+2V'\cos B}{b}\beta$$
$$+\frac{U+V+2W'\cos C}{c}\gamma=0.$$

21. *To find the magnitude of the axes of the conic.*

Let $\bar{\alpha}, \bar{\beta}, \bar{\gamma}$ be the co-ordinates of the centre; and, for shortness' sake, put

$$\alpha-\bar{\alpha}=x, \quad \beta-\bar{\beta}=y, \quad \gamma-\bar{\gamma}=z.$$

Then if r be the semi-diameter drawn from the centre to α, β, γ, we have (see Art. 3, Chap. I.)

$$r^2=\frac{abc}{4\Delta^2}(a\cos A.x^2+b\cos B.y^2+c\cos C.z^2)\ldots\ldots(1).$$

Again, from the equation of the conic,

$$0=\phi(\alpha,\beta,\gamma)=\phi(\bar{\alpha}+x,\bar{\beta}+y,\bar{\gamma}+z)$$
$$=\phi(\bar{\alpha},\bar{\beta},\bar{\gamma})+2x(u\bar{\alpha}+w'\bar{\beta}+v'\bar{\gamma})$$
$$+2y(w'\bar{\alpha}+v\bar{\beta}+u'\bar{\gamma})+2z(v'\bar{\alpha}+u'\bar{\beta}+w\bar{\gamma})$$
$$+\phi(x,y,z).$$

Now, by Art. 11 of the present chapter,

$$\frac{u\bar{\alpha}+w'\bar{\beta}+v'\bar{\gamma}}{a}=\frac{w'\bar{\alpha}+v\bar{\beta}+u'\bar{\gamma}}{b}=\frac{v'\bar{\alpha}+u'\bar{\beta}+w\bar{\gamma}}{c}.$$

MAGNITUDES OF THE AXES OF THE CONIC.

Also, $ax + by + cz = a(\alpha - \bar{\alpha}) + b(\beta - \bar{\beta}) + c(\gamma - \bar{\gamma}) = 0 \ldots (2)$;

$\therefore \phi(x, y, z) = -\phi(\bar{\alpha}, \bar{\beta}, \bar{\gamma})$,

or, $ux^2 + vy^2 + wz^2 + 2u'yz + 2v'zx + 2w'xy$

$$= \frac{\begin{vmatrix} u, & w', & v' \\ w', & v, & u' \\ v', & u', & w \end{vmatrix}(2\Delta)^2}{\begin{vmatrix} u, & w', & v', & a \\ w', & v, & u', & b \\ v', & u', & w, & c \\ a, & b, & c, & 0 \end{vmatrix}} \ldots\ldots\ldots\ldots\ldots(3).$$

(See Art. 12, Cor.)

Now the semi-axes are the greatest and least values of the semi-diameter. We have then to make

$$\frac{4\Delta^2}{abc} r^2 = a\cos A \cdot x^2 + b\cos B \cdot y^2 + c\cos C \cdot z^2 \ldots\ldots(4)$$

a maximum or minimum, x, y, z being connected by the relations (2) and (3).

Multiply (2) by the indeterminate multiplier 2μ, (4) by λ, adding them to (3), differentiating, and equating to zero the coefficients of each differential, we get

$$\left. \begin{array}{l} ux + w'y + v'z + \lambda a\cos A \cdot x + \mu a = 0 \\ w'x + vy + u'z + \lambda b\cos B \cdot y + \mu b = 0 \\ v'x + u'y + wz + \lambda c\cos C \cdot z + \mu c = 0 \end{array} \right\} \ldots\ldots\ldots(5).$$

Multiplying these equations in order by x, y, z, and adding, we get

$$\frac{\begin{vmatrix} u, & w', & v' \\ w', & v, & u' \\ v', & u', & w \end{vmatrix}}{\begin{vmatrix} u, & w', & v', & a \\ w', & v, & u', & b \\ v', & u', & w, & c \\ a, & b, & c, & 0 \end{vmatrix}} + \lambda \frac{r^2}{abc} = 0.$$

TRILINEAR CO-ORDINATES.

Substituting this value of λ in equations (5), and eliminating x, y, z, μ from the equations combined with (2), we obtain the following quadratic for the determination of $\frac{1}{r^2}$:

$$\begin{vmatrix} \left(\dfrac{as\cos A}{r^2} - u\right), & -w', & -v', & a \\ -w', & \left(\dfrac{bs\cos B}{r^2} - v\right), & -u', & b \\ -v', & -u', & \left(\dfrac{cs\cos C}{r^2} - w\right), & c \\ a, & b, & c, & 0 \end{vmatrix} = 0,$$

where s is written for
$$\dfrac{abc \begin{vmatrix} u, & w', & v' \\ w', & v, & u' \\ v', & u', & w \end{vmatrix}}{\begin{vmatrix} u, & w', & v', & a \\ w', & v, & u', & b \\ v', & u', & w, & c \\ a, & b, & c, & 0 \end{vmatrix}}.$$

This equation determines the semi-axes.

22. *To find the area of the conic.*

In the above equation, the coefficient of $\frac{1}{r^2}$ is

$$-abcs^2(a\cos B\cos C + b\cos C\cos A + c\cos A\cos B),$$

which is equal to

$$-\frac{a^2b^2c^2s^2}{2\Delta}(\sin A\cos B\cos C + \sin B\cos C\cos A$$
$$+ \sin C\cos A\cos B)$$
$$= -\frac{a^2b^2c^2s^2}{2\Delta}\sin A\sin B\sin C = -4\Delta^2 \cdot s^2.$$

The term independent of r^2 is

$$\begin{vmatrix} u, & w', & v', & a \\ w', & v, & u', & b \\ v', & u', & w, & c \\ a, & b, & c, & 0 \end{vmatrix}$$

Hence the product of the two values of r^2 is

$$-\frac{4\Delta^2 s^2}{\begin{vmatrix} u, & w', & v', & a \\ w', & v, & u', & b \\ v', & u', & w, & c \\ a, & b, & c, & 0 \end{vmatrix}}.$$

The area of the conic is, therefore,

$$2\pi\Delta abc \frac{\begin{vmatrix} u, & w', & v' \\ w', & v, & u' \\ v', & u', & w \end{vmatrix}}{\begin{vmatrix} u, & w', & v', & -a \\ w', & v, & u', & -b \\ v', & u', & w, & -c \\ c, & b, & c, & 0 \end{vmatrix}^{\frac{3}{2}}}.$$

From the above investigation may be obtained the criterion which determines whether the conic be an ellipse or hyperbola. For, in the hyperbola, the two values of r^2 have opposite signs, hence the curve will be an ellipse or hyperbola according as

$$\begin{vmatrix} u, & w', & v', & a \\ w', & v, & u', & b \\ v', & u', & w, & c \\ a, & b, & c, & 0 \end{vmatrix}$$

is negative or positive; or according as

$$Ua^2 + Vb^2 + Wc^2 + 2U'bc + 2V'ca + 2W'ab$$

is positive or negative.

F.

EXAMPLES.

1. Each angular point of a triangle is joined with each of two given points; prove that the six points of intersection of the joining lines with the opposite sides of the triangle lie in a conic.

2. A conic is described, touching three given straight lines and passing through a given point; prove that the locus of it centre is a conic.

Express, in geometrical language, the position of the given point relatively to the straight lines, in order that the locus of the centre may be a circle.

Also find the locus of the given point, in order that the locus of the centre may be a rectangular hyperbola.

3. Four circles are described, so that each of the four triangles, formed by each three of four given straight lines, is self-conjugate with respect to one of them; prove that the four circles have a common radical axis.

4. If A, B, C, A', B', C' be six points, such that the straight lines $B'C'$, $C'A'$, $A'B'$, are the several polars of the points A, B, C, with respect to a given conic, prove that

The three straight lines AA', BB', CC', intersect in a point; and that

The points of intersection of BC with $B'C'$, CA with $C'A'$, AB with $A'B'$, lie in a straight line.

5. If two triangles circumscribe a conic, their angular points lie on another conic.

6. The equation of a conic circumscribing the triangle of reference, and having its semi-diameters parallel to the sides equal to r_1, r_2, r_3, respectively, is

$$\frac{a}{r_1^2 \alpha} + \frac{b}{r_2^2 \beta} + \frac{c}{r_3^2 \gamma} = 0.$$

EXAMPLES.

7. A conic always touches the sides of a given triangle; prove that, if the sum of the squares on its axes be given, the locus of its centre is a circle, the centre of which is the point of intersection of the perpendiculars let fall from the angular points of the triangle on the opposite sides.

8. If θ be the angle between the asymptotes of the conic, represented by the general equation of the second degree, prove that

$$4 \begin{vmatrix} 0, & \sin A, & \sin B, & \sin C \\ \sin A, & u, & w', & v' \\ \sin B, & w', & v, & u' \\ \sin C, & v', & u', & w \end{vmatrix} - (u + v + w - 2u'\cos A - 2v'\cos B - 2w'\cos C)^2 \tan^2\theta = 0.$$

9. The two circular points at infinity may be represented by the equations,

$$-a = \beta \epsilon^{-\sqrt{-1}C} = \gamma \epsilon^{\sqrt{-1}B},$$

$$-a = \beta \epsilon^{\sqrt{-1}C} = \gamma \epsilon^{-\sqrt{-1}B}.$$

10. Prove that the area of a conic inscribed in the triangle of reference is to the area of the triangle as 2π is to $\dfrac{a}{d_1} + \dfrac{b}{d_2} + \dfrac{c}{d_3}$, where d_1, d_2, d_3 are the lengths of the semi-diameters parallel to a, b, c respectively.

11. The internal and external bisectors of the angle BAC of the triangle of reference meet the side BC in L_1, L_2, respectively; M_1, M_2, N_1, N_2 are points similarly taken on the sides CA, AB respectively: circles are described on L_1L_2, M_1M_2, N_1N_2 as diameters; prove that these circles have a common radical axis, whose equation is

$$(b^2 - c^2) bc\alpha + (c^2 - a^2) ca\beta + (a^2 - b^2) ab\gamma = 0.$$

12. Prove that the circumscribed circle, the self-conjugate circle, and the nine-point circle of a triangle, have a common radical axis.

13. Prove that twice the square on the tangent, drawn from any external point to the nine-point circle, is equal to the sum of the squares on the tangents, drawn from the same point to the circumscribed and self-conjugate circles.

CHAPTER V.

TRIANGULAR CO-ORDINATES.

1. WE shall now give a concise account of a system of co-ordinates which differs from that which has been the subject of the preceding chapters in assigning a slightly different interpretation to the co-ordinates. In the system which we are about to explain, the position of a point P is considered as determined by the ratios of the *areas of the triangles* PBC, PCA, PAB, to the triangle of reference ABC. If these quantities be denoted by the letters x, y, z, they will be connected by the identical relation

$$x + y + z = 1.$$

2. In this method, as in that of trilinear co-ordinates, an equation of the first degree represents a straight line, and one of the second degree a conic.

Again, since $x : a\alpha :: y : b\beta :: z : c\gamma$, it follows that if the same straight line be represented in the two systems by the equations

$$l\alpha + m\beta + n\gamma = 0,$$
$$l'x + m'y + n'z = 0;$$
$$\therefore l : l'a :: m : m'b :: n : n'c.$$

Hence we may pass from any relation among the coefficients in the trilinear system to that in the present one, by writing la, mb, nc, for l, m, n, respectively. Similarly, in conics, we may pass from any such formula to the corresponding one, by writing

$$ua^2, vb^2, wc^2, u'bc, v'ca, w'ab, \text{ for } u, v, w, u', v', w'.$$

And, since $U = vw - u'^2$,

we must write for U, b^2c^2U, and similarly for V and W, c^2a^2V, a^2b^2W.

Also, since $U' = v'w' - uu'$,

we must write for U', a^2bcU', and similarly for V' and W', b^2caV', c^2abW''.

Hence we obtain the following synopsis of formulæ:

The straight lines drawn through the angular points of a triangle, bisecting the opposite sides, are represented by

$$y - z = 0, \quad z - x = 0, \quad x - y = 0.$$

The internal bisectors of the angles, by

$$\frac{y}{b} - \frac{z}{c} = 0, \quad \frac{z}{c} - \frac{x}{a} = 0, \quad \frac{x}{a} - \frac{y}{b} = 0.$$

The perpendiculars, by

$$y \cot B - z \cot C = 0, \quad z \cot C - x \cot A = 0,$$
$$x \cot A - y \cot B = 0.$$

The distance between two points, by

$$\{a^2(y - y')(z' - z) + b^2(z - z')(x' - x) + c^2(x - x')(y' - y)\}^{\frac{1}{2}}$$

or by

$$\left\{\frac{(b^2 + c^2 - a^2)(x - x')^2 + (c^2 + a^2 - b^2)(y - y')^2 + (a^2 + b^2 - c^2)(z - z')^2}{2}\right\}^{\frac{1}{2}}.$$

The condition of parallelism of the straight lines

$$lx + my + nz = 0, \quad l'x + m'y + n'z = 0, \text{ is}$$

$$\begin{vmatrix} 1, & l, & l' \\ 1, & m, & m' \\ 1, & n, & n' \end{vmatrix} = 0,$$

or $mn' - m'n + nl' - n'l + lm' - l'm = 0$.

The condition of perpendicularity,

$$2ll'a^2 + 2mm'b^2 + 2nn'c^2 - (mn' + m'n)(b^2 + c^2 - a^2)$$
$$- (nl' + n'l)(c^2 + a^2 - b^2)$$
$$- (lm' + l'm)(a^2 + b^2 - c^2) = 0,$$

or $\{(l-m)(l'-n') + (l-n)(l'-m')\} a^2$
$+ \{(m-n)(m'-l') + (m-l)(m'-n')\} b^2$
$+ \{(n-l)(n'-m') + (n-m)(n'-l')\} c^2 = 0.$

The perpendicular distance from the point (x, y, z) to the line $lx + my + nz = 0$, is

$$\frac{(lx + my + nz) 2\Delta}{\{(l-m)(l-n) a^2 + (m-n)(m-l) b^2 + (n-l)(n-m) c^2\}^{\frac{1}{2}}}.$$

The line at infinity will be represented by $x + y + z = 0$.

3. Again, in conics we have the following formulæ:

The conic will be a parabola, if

$$\begin{vmatrix} u, & w', & v', & 1 \\ w', & v, & u', & 1 \\ v', & u', & w, & 1 \\ 1, & 1, & 1, & 0 \end{vmatrix} = 0,$$

or if $\quad U + V + W + 2U' + 2V' + 2W' = 0.$

A rectangular hyperbola, if

$$ua^2 + vb^2 + wc^2 - u'(b^2 + c^2 - a^2) - v'(c^2 + a^2 - b^2)$$
$$- w'(a^2 + b^2 - c^2) = 0,$$

or $\quad (u + u' - v' - w') a^2 + (v + v' - w' - u') b^2$
$\qquad + (w + w' - u' - v') c^2 = 0.$

A circle, if

$$\frac{v + w - 2u'}{a^2} = \frac{w + u - 2v'}{b^2} = \frac{u + v - 2w'}{c^2}.$$

The centre is given by

$$\frac{\bar{x}}{\begin{vmatrix} w', & v', & 1 \\ v, & u', & 1 \\ u', & w, & 1 \end{vmatrix}} = \frac{\bar{y}}{\begin{vmatrix} u', & w', & 1 \\ w, & v', & 1 \\ v', & u, & 1 \end{vmatrix}} = \frac{\bar{z}}{\begin{vmatrix} v', & u', & 1 \\ u, & w', & 1 \\ w', & v, & 1 \end{vmatrix}}$$

or $\dfrac{\bar{x}}{U+V'+W'} = \dfrac{\bar{y}}{V+W'+U'} = \dfrac{\bar{z}}{W+U'+V'}$.

The equation of the asymptotes is

$$\begin{vmatrix} u, & w', & v', & 1 \\ w', & v, & u', & 1 \\ v', & u', & w, & 1 \\ 1, & 1, & 1, & 0 \end{vmatrix} \phi(x, y, z) + \begin{vmatrix} u, & w', & v' \\ w', & v, & u' \\ v', & u', & w \end{vmatrix} (x+y+z)^2 = 0.$$

The radical axis of two circles

$$ux^2 + vy^2 + wz^2 + 2u'yz + 2v'zx + 2w'xy = 0,$$
$$px^2 + qy^2 + rz^2 + 2p'yz + 2q'zx + 2r'xy = 0,$$

is represented by the equation

$$\frac{ux+vy+wz}{u+v+w-u'-v'-w'} = \frac{px+qy+rz}{p+q+r-p'-q'-r'}.$$

The circular points at infinity by

$$-\frac{x}{a} = \epsilon^{\pm\sqrt{-1}B} \frac{y}{b} = \epsilon^{\pm\sqrt{-1}C} \frac{z}{c}.$$

The inscribed circle by

$$(s-a)^2 x^2 + (s-b)^2 y^2 + (s-c)^2 z^2$$
$$- 2(s-b)(s-c)yz - 2(s-c)(s-a)zx - 2(s-a)(s-b)xy = 0.$$

The square on the tangent from the point x, y, z to the circle $ux^2+\ldots+2u'yz+\ldots=0$ is

$$\tfrac{1}{2} \frac{a^2+b^2+c^2}{u+v+w-u'-v'-w'} (ux^2+vy^2+wz^2+2u'yz+2v'zx+2w'xy).$$

Other formulæ may be adapted in a similar manner.

CHAPTER VI.

RECIPROCAL POLARS.

1. The theory of Reciprocal Polars, which will be treated of in this chapter, discusses the relations which exist between systems of points and straight lines which are the poles and polars of each other with regard to any conic; and shews how from the properties of a curve, regarded as the locus of a moving point, may be deduced those of another curve which is always touched by the polar of this moving point with regard to a fixed conic. The theory is especially valuable when the conic, with respect to which the poles and polars are taken, is a circle.

2. The polar of the point of intersection of two given straight lines is the straight line which joins the poles of those straight lines. This will readily be seen to follow geometrically from the definitions of a pole and polar; or it may be analytically proved thus.

Let the two straight lines be represented by the equations

$$l_1\alpha + m_1\beta + n_1\gamma = 0 \quad \ldots\ldots\ldots\ldots\ldots (1),$$
$$l_2\alpha + m_2\beta + n_2\gamma = 0 \quad \ldots\ldots\ldots\ldots\ldots (2).$$

At their point of intersection, we have

$$\frac{\alpha}{m_1 n_2 - m_2 n_1} = \frac{\beta}{n_1 l_2 - n_2 l_1} = \frac{\gamma}{l_1 m_2 - l_2 m_1}.$$

The polar of this with respect to

$$L\alpha^2 + M\beta^2 + N\gamma^2 = 0,$$

to which form every conic may, by suitable choice of the triangle of reference, be reduced, is represented by the equation

$$(m_1 n_2 - m_2 n_1) L\alpha + (n_1 l_2 - n_2 l_1) M\beta + (l_1 m_2 - l_2 m_1) N\gamma = 0 \ldots (3).$$

But the poles of (1) and (2) with respect to the same conic are given by

$$\frac{L\alpha}{l_1} = \frac{M\beta}{m_1} = \frac{N\gamma}{n_1},$$

$$\frac{L\alpha}{l_2} = \frac{M\beta}{m_2} = \frac{N\gamma}{n_2}.$$

Both these points lie on the line (3). Hence the proposition is proved.

3. If a point move in any manner whatever, its polar will move in a manner dependent upon the motion of the point, and the curve which the polar always touches (its *envelope*, as it is called) will have certain definite relations to the path traced out by the point. The locus of the moving point and the envelope of its polar, are called the *polar reciprocals* of one another. The use of the word *reciprocal* arises from the fact, which we proceed to demonstrate, that the locus of the point may be generated from the envelope of its polar, in the same manner as the latter curve was generated from the former. For shortness' sake we shall denote the two curves by the letters L and E.

Let P, P' be any two points on L, the pole Q of the chord PP' will be the point of intersection of the corresponding tangents to E (that is, of the two tangents to E which are the polars of P, P' with respect to the conic). Now let P' move along L up to P, then PP' ultimately becomes the tangent to L at P; moreover the polars of P and P' approach indefinitely near to coincidence, and their point of intersection Q will ultimately be a point on E. But Q is the pole of PP', hence *the polar of any point on E is a tangent to L.* That is, if a point move along E, its polar will envelope L. In other words, L may be generated from E, as E was from L. In this consists the reciprocity of the curves.

The process of generating E from L, or L from E, is called *reciprocating L or E.*

4. If the curve L be cut by any straight line whatever, the polars of the several points of intersection will be the several tangents of E, drawn through the pole of the cutting line. And conversely, the several tangents drawn to L from any point will have for their poles the several points in which E is intersected by the polar of that point.

If any two curves be reciprocated, the polar of any point common to both will be a common tangent to the reciprocal curves, and the pole of any tangent common to both will be a point of intersection of the reciprocal curves. Hence any two curves will have as many points of intersection as their reciprocals have common tangents, and as many common tangents as their reciprocals have points of intersection.

If the curves touch one another, then two of their points of intersection coincide; and consequently the two corresponding tangents to the reciprocal curves will coincide, and therefore the reciprocal curves will also touch one another.

5. From what has been said above, it will be seen that the total number of tangents, real or imaginary, which can be drawn to E or L from any point (not on the curve itself) is equal to the total number of points, real or imaginary, in which L or E is cut by any straight line, not a tangent to it.

A curve, to which n tangents can be drawn through the same point, is said to be of the nth *class*, and we may therefore express the above proposition by saying that the degree of a curve is the same as the class of its reciprocal, and the class of a curve the same as the degree of its reciprocal.

6. The number of tangents, real or imaginary, which can be drawn to a conic from a given point is known to be two. Hence, the reciprocal of a conic is intersected by a given straight line in two points, real or imaginary, and is therefore of the second degree, that is, it is itself a conic.

7. This proposition may also be proved analytically as follows.

DEF. The conic with respect to which the poles and polars are taken is called the *auxiliary* conic.

We have seen (Art. 15, Chap. II.) that any two conics may be expressed by equations involving the squares of the variables only. Let then the auxiliary conic be denoted by

$$L\alpha^2 + M\beta^2 + N\gamma^2 = 0 \dots\dots\dots\dots\dots(1),$$

and the conic to be reciprocated by

$$l\alpha^2 + m\beta^2 + n\gamma^2 = 0 \dots\dots\dots\dots\dots(2).$$

If (f, g, h) be any point on the required curve, its polar with respect to (1) will be given by the equation

$$Lf\alpha + Mg\beta + Nh\gamma = 0.$$

In order that this may touch (2) we must have (see Art. 16, Chap. II.)

$$\frac{L^2}{l}f^2 + \frac{M^2}{m}g^2 + \frac{N^2}{n}h^2 = 0 \dots\dots\dots\dots(3).$$

(3), regarding f, g, h as current co-ordinates, is therefore the reciprocal of (2) with respect to (1).

COR. It hence appears that the three points which form a conjugate triad for two given conics, will also form a conjugate triad for the reciprocal of one with respect to the other.

8. *To find the polar reciprocal of the conic*

$$u\alpha^2 + v\beta^2 + w\gamma^2 + 2u'\beta\gamma + 2v'\gamma\alpha + 2w'\alpha\beta = 0$$

with respect to

$$\alpha^2 + \beta^2 + \gamma^2 = 0.$$

[The conic, $\alpha^2 + \beta^2 + \gamma^2 = 0$, is imaginary, but the analytical process of finding the pole of a given straight line, or the polar of a given point, may be equally well performed, whether the auxiliary conic be imaginary or real, provided its coefficients be real.]

Let f, g, h be any point on the required locus, its polar with respect to the auxiliary conic is

$$f\alpha + g\beta + h\gamma = 0,$$

and in order that this may touch the given conic, we must have (Art. 4, Chap. IV.)

$$\begin{vmatrix} 0, & f, & g, & h \\ f, & u, & w', & v' \\ g, & w', & v, & u' \\ h, & v', & u', & w \end{vmatrix} = 0,$$

or, $(vw - u'^2)f^2 + (wu - v'^2)g^2 + (uv - w'^2)h^2 + 2(v'w' - uu')gh + 2(w'u' - vv')hf + 2(u'v' - ww')fg = 0,$

which, adopting the notation of Chap. IV., may be written

$$Uf^2 + Vg^2 + Wh^2 + 2U'gh + 2V'hf + 2W'fg = 0.$$

This is therefore the required equation.

It may be proved, in a similar manner, that if

$$\phi(\alpha, \beta, \gamma) = 0, \quad \psi(\alpha, \beta, \gamma) = 0$$

be the equations of any two conics, the equation of the reciprocal of the first with respect to the second is

$$\begin{vmatrix} 0, & \dfrac{d\psi}{d\alpha}, & \dfrac{d\psi}{d\beta}, & \dfrac{d\psi}{d\gamma} \\ \dfrac{d\psi}{d\alpha}, & \dfrac{d^2\phi}{d\alpha^2}, & \dfrac{d^2\phi}{d\alpha d\beta}, & \dfrac{d^2\phi}{d\alpha d\gamma} \\ \dfrac{d\psi}{d\beta}, & \dfrac{d^2\phi}{d\beta d\alpha}, & \dfrac{d^2\phi}{d\beta^2}, & \dfrac{d^2\phi}{d\beta d\gamma} \\ \dfrac{d\psi}{d\gamma}, & \dfrac{d^2\phi}{d\gamma d\alpha}, & \dfrac{d^2\phi}{d\gamma d\beta}, & \dfrac{d^2\phi}{d\gamma^2} \end{vmatrix} = 0.$$

9. Since the tangents at the extremity of any diameter of a conic are parallel to one another, it follows that the polar of the centre is at an infinite distance, and conversely, that the line at infinity reciprocates into the centre of the auxiliary conic. Hence it follows that parallel lines reciprocate into points lying on a straight line passing through the centre of the auxiliary conic; and that the asymptotes of any curve, being the tangents drawn to it at the points where it meets the line at infinity, reciprocate into the points of contact of the tangents drawn to the reciprocal curve from the centre of the auxiliary conic.

Since the asymptotes of an hyperbola are real, while those of an ellipse are imaginary, it follows that the tangents, drawn from the centre of the auxiliary conic (supposed real) to the reciprocal curve, will be real or imaginary, according as the original curve is an hyperbola or an ellipse. If it be a parabola, the reciprocal curve will pass through the centre of the conic, which is in accordance with what has already been stated, that every parabola touches the line at infinity. Conversely, if one conic be reciprocated with respect to another, the reciprocal curve will be an ellipse, parabola, or hyperbola, according as the centre of the auxiliary conic lies within, upon, or without, the original conic.

10. We have now sufficient materials for transforming any *descriptive* proposition, that is any proposition relating to the position of lines and points, without reference to considerations of magnitude, into another. Before proceeding further, we will give a few examples of this process.

We will first take the following proposition. "If two of the angular points of a triangle move each along a fixed straight line, and each side pass through a fixed point, the three points lying in the same straight line, the third angular point will move along a straight line, passing through the intersection of the straight lines along which the other angular points move."

The reciprocals of the three sides of the given triangle will be three points, which may be considered as the angles of a triangle, which may be called the reciprocal triangle. Those of the angular points of the first triangle will be the sides of the reciprocal. Those of the fixed straight lines, along which two of the angular points of the first triangle move, will be fixed points through which two of the sides of the reciprocal triangle pass. Those of the three points, lying in the same straight line, through which the sides of the given triangle always pass, will be three straight lines, intersecting in a point, along which the angular points of the reciprocal triangle always move. Hence the data of the reciprocal proposition will be "Two of the sides of a triangle pass each through a fixed point, and each angular point moves along a fixed straight line, the three straight lines passing through the same point." In the given theorem, the thing to be proved relates to the motion of the third angular point. To this will correspond the third side of the reciprocal triangle. To the straight line, passing through the intersection of the two given straight lines, along which the third angular point may be shewn to move, corresponds a point lying in the same straight line with the two given points, and through this the third side will always pass. Hence, under the circumstances stated above as data of the reciprocal theorem, "the third side will pass through a fixed point lying in the straight line joining the two fixed points, through which the first sides pass*."

* The given theorem may be expressed, by the aid of letters, as follows:

Let PQR be the given triangle, and let its angular point Q move along a fixed straight line OX, its angular point R along a fixed straight line OY. Also, let the straight line QR always pass through a fixed point F, RP through a fixed point G, PQ through a fixed point H, the three points F, G, H lying in the same straight line. Then the given theorem tells us that the point P will always move along a fixed straight line, passing through O.

Now let the whole figure be reciprocated with respect to any conic section. Let the line which is the polar of any point be denoted by accenting the same *single letter* by which the point is denoted in the original figure; the polar of P, for example, being denoted by P'. Then the *point of intersection* of the lines P', Q' will be denoted by the *two letters* $P'Q'$, and this will be the pole of the line PQ. We have then a triangle of which the sides are P', Q', R', the side Q' always passing through a fixed point $O'X'$, the side R' through a fixed point $O'Y'$. Also the angular point $Q'R'$ always moves along a fixed straight line F', the point $R'P'$ along a fixed straight line G', the point $P'Q'$ along a fixed straight line H', the three straight lines F', G', H' passing through the same point. Then the reciprocal theorem is that the side P' will always pass through a fixed point lying in the line Q'.

The student will find the above mode of transformation, in which a

Again, turn to Example 4, on page 56, and let us investigate the reciprocal theorem. The three conics touching respectively each pair of the sides of a triangle at the angular points where they meet the third side, will reciprocate into "three conics passing respectively through each pair of the angular points of a triangle, and touching the lines joining them with the third angular point," that is, the sides of the triangle themselves. This condition, therefore, reciprocates into itself. The condition "all intersecting in a point" reciprocates into "all touching a straight line." Hence the data are, "Three conics are drawn, touching respectively each pair of the angular points of the sides of a triangle at the points where they meet the third side, and all touching a straight line."

In the matter to be proved, we may first enquire what are the reciprocals of "the sides of the triangle which intersect" (that is, which do not touch) "their respective conics." These will be "the angular points of the triangle not lying on their respective conics." The three tangents at their common point will reciprocate into "the three points of contact of their common tangents." And the meeting of the tangents with the sides will reciprocate into the lines joining the points of contact with the angular points. Hence the first thing to be proved is, "That the three straight lines joining the points of contact of the common tangent with the angular points of the triangle not lying on the respective conics all pass through a point."

Again, "the other common tangents to each pair of conics" reciprocate into "the other points of intersection of each pair of conics," and "the sides of the triangle which touch the several pairs of conics" into the angular points of the triangle "common to the several pairs of conics." Hence the latter part of the theorem will run: "And that the same three straight lines respectively join the other point of intersection of each pair of conics with the angular point of the triangle common to each pair."

straight line is denoted by a single letter, and a point by the pair of letters representing any two straight lines which intersect in it, a useful mode of familiarizing himself with the method of reciprocal polars.

11. After a little practice, the process of reciprocating a given theorem will be found to consist simply in writing "straight line" for "point," "join" for "intersect," "locus" for "envelope," &c., and *vice versâ*. The word "conic" will of course remain unaltered.

12. *Brianchon's Theorem.*

By reciprocating Paschal's Theorem (given in Art. 12, Chap. III.), we obtain Brianchon's Theorem, which asserts that

"If a hexagon be described about a conic section, the three diagonals will intersect in a point*."

---

* It may be well to append an independent proof of this important theorem.

Take three sides of the hexagon as lines of reference, and let the equations of the other three be

$$\alpha + m_1\beta + n_1\gamma = 0, \quad l_2\alpha + \beta + n_2\gamma = 0, \quad l_3\alpha + m_3\beta + \gamma = 0.$$

Let the equation of the conic be

$$(L\alpha)^{\frac{1}{2}} + (M\beta)^{\frac{1}{2}} + (N\gamma)^{\frac{1}{2}} = 0.$$

The conditions of tangency are

$$L + \frac{M}{m_1} + \frac{N}{n_1} = 0,$$

$$\frac{L}{l_2} + M + \frac{N}{n_2} = 0,$$

$$\frac{L}{l_3} + \frac{M}{m_3} + N = 0;$$

whence

$$\begin{vmatrix} 1, & \dfrac{1}{m_1}, & \dfrac{1}{n_1} \\ \dfrac{1}{l_2}, & 1, & \dfrac{1}{n_2} \\ \dfrac{1}{l_3}, & \dfrac{1}{m_3}, & 1 \end{vmatrix} = 0.$$

The line passing through the intersections of $\beta = 0$ with $(l_3, m_3, 1)$ and of $\gamma = 0$ with $(l_2, 1, n_2)$ is represented by the equation

$$\alpha + \frac{\beta}{l_2} + \frac{\gamma}{l_3} = 0.$$

Similarly, the other two diagonals are represented by

$$\frac{\alpha}{m_1} + \beta + \frac{\gamma}{m_3} = 0,$$

$$\frac{\alpha}{n_1} + \frac{\beta}{n_2} + \gamma = 0;$$

The student will find it useful to transform, by the method of reciprocal polars, the special cases of Pascal's Theorem, given in Art. 13, Chap. III.; and to obtain a geometrical construction by which *when five tangents to a conic are given, their points of contact may be found.*

13. The anharmonic ratio of the pencil formed by four intersecting straight lines is the same as that of the range formed by their poles. This may be proved as follows.

Let OP, OQ, OR, OS be the four straight lines, P', Q', R', S' their poles, which will lie in a straight line, the polar

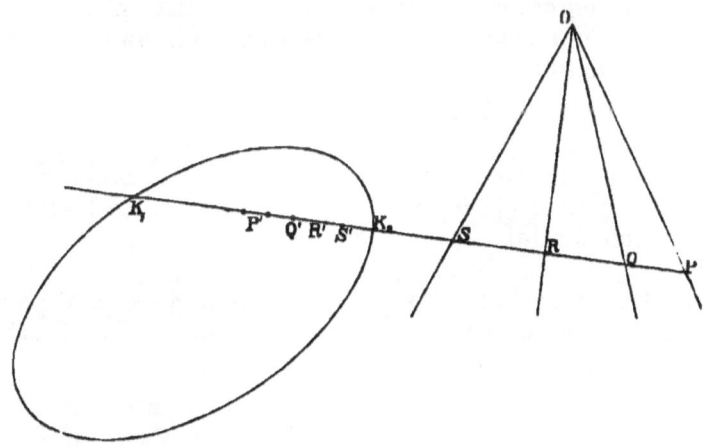

Fig. 18.

of O; let P, Q, R, S be the points in which the pencil is cut by the transversal $P'Q'R'S'$.

Let this transversal cut the conic in K_1, K_2. Bisect

whence, if these intersect in a point,
$$\begin{vmatrix} 1, & \dfrac{1}{l_2}, & \dfrac{1}{l_3} \\ \dfrac{1}{m_1}, & 1, & \dfrac{1}{m_3} \\ \dfrac{1}{n_1}, & \dfrac{1}{n_2}, & 1 \end{vmatrix} = 0;$$

the same condition as that already investigated.

F.

$K_1 K_2$ in V. Then, since $PK_1P'K_2$ is divided harmonically in P, K_1, P', K_2 (Art. 21, Chap. II.), it follows that

$$PK_1 . P'K_2 = PK_2 . P'K_1,$$

whence

$$(VP - VK_1)(VK_2 + VP') = (VP + VK_2)(VK_1 - VP'),$$

which since $VK_1 = VK_2$, reduces to

$$VP . VP' = VK_1^2,$$

which, by similar reasoning,

$$= VQ . VQ' = VR . VR' = VS . VS'.$$

Hence the eight points P, Q, R, S, P', Q', R', S' are in an involution, of which K_1, K_2 are the foci, and therefore (Art. 27, Chap. I.)

$$\{O . PQRS\} = [P'Q'R'S'].$$

14. In Art. 13, Chap. I. we saw that the condition that the three points (l_1, m_1, n_1), (l_2, m_2, n_2), (l_3, m_3, n_3) shall lie in the same straight line is identical with the condition that the three straight lines (l_1, m_1, n_1), (l_2, m_2, n_2), (l_3, m_3, n_3) shall intersect in the same point. Now these several points and lines are respectively the poles and polars of each other, with respect to the imaginary conic

$$\alpha^2 + \beta^2 + \gamma^2 = 0.$$

Thus the theory of reciprocal polars explains the fact that the condition for three points lying in a straight line is identical with that for three straight lines insersecting in a point. It also explains the identity of conditions noticed in Chap. II. Arts. 7 and 9.

For the reciprocal of the conic

$$\lambda^2\alpha^2 + \mu^2\beta^2 + \nu^2\gamma^2 - 2\mu\nu\beta\gamma - 2\nu\lambda\gamma\alpha - 2\lambda\mu\alpha\beta = 0 \ldots\ldots (1),$$

with respect to

$$\alpha^2 + \beta^2 + \gamma^2 = 0,$$

will be found to be

$$\lambda\beta\gamma + \mu\gamma\alpha + \nu\alpha\beta = 0 \ldots\ldots\ldots\ldots\ldots (2).$$

And the polar of (f, g, h) is $f\alpha + g\beta + h\gamma = 0$.

Hence if the line $f\alpha + g\beta + h\gamma = 0$ touch (1), the point (f, g, h) lies in (2), giving for the condition of tangency

$$\frac{\lambda}{f} + \frac{\mu}{g} + \frac{\nu}{h} = 0.$$

And if the line $f\alpha + g\beta + h\gamma = 0$ touch (2), the point (f, g, h) lies in (1), giving for the condition of tangency in that case

$$\lambda^2 f^2 + \mu^2 g^2 + \nu^2 h^2 - 2\mu\nu gh - 2\nu\lambda hf - 2\lambda\mu fg = 0.$$

These conditions of tangency are identical with those already investigated.

Again, every parabola touches the line at infinity. Now the co-ordinates of the pole of this line are proportional to a, b, c. Hence, if the conic, represented by the general equation of the second degree, be a parabola, the point (a, b, c) must lie in the reciprocal conic. This gives, as the condition for a parabola,

$$Ua^2 + Vb^2 + Wc^2 + 2U'bc + 2V'ca + 2W'ab = 0,$$

the same as that already investigated.

15. Prop. *Any straight line drawn through a given plane A is divided harmonically by any conic section, and the polar of A with respect to it.*

This proposition may be proved as follows. Let the straight line cut the curve in P and Q, and the polar of A in B. Let C be the polar of the straight line, and let ABC be the triangle of reference. The conic will be self-conjugate with respect to ABC, and will therefore be represented by the equation

$$u\alpha^2 + v\beta^2 + w\gamma^2 = 0.$$

Hence the lines CP, CQ, which are tangents to the conic, are represented by the equation

$$u\alpha^2 + v\beta^2 = 0,$$

and therefore form an harmonic pencil with CA, CB.

116 TRILINEAR CO-ORDINATES.

16. The straight line CB may be regarded as the polar of A with respect to the locus made up of the two straight lines CP, CQ. For the values of β and γ at A being $0, 0$, and the equation of CP, CQ being $u\alpha^2 + v\beta^2 = 0$, we get for the equation of its polar, $\alpha = 0$, that is, the polar is the line AB.

17. If four straight lines form an harmonic pencil, either pair will be its own polar reciprocal with respect to the other. For, adapting the equation of Art. 8, to the case of two variables only, we get for the polar reciprocal of $\alpha\beta = 0$, with respect to $u\alpha^2 + v\beta^2 = 0$, the following equation,

$$\begin{vmatrix} 0, & u\alpha, & v\beta \\ u\alpha, & 0, & 1 \\ v\beta, & 1, & 0 \end{vmatrix} = 0,$$

or $uv\alpha\beta = 0$.

And, conversely, for that of $u\alpha^2 + v\beta^2 = 0$ with respect to $\alpha\beta = 0$,

$$\begin{vmatrix} 0, & \beta, & \alpha \\ \beta, & u, & 0 \\ \alpha, & 0, & v \end{vmatrix} = 0,$$

or $u\alpha^2 + v\beta^2 = 0$, in either case reproducing the reciprocated curve. Hence the proposition is proved.

18. We may hence deduce the condition that two pairs of straight lines may form an harmonic pencil. First let them all intersect in A, and the equations of the two pairs be

$$u\alpha^2 + v\beta^2 + 2w'\alpha\beta = 0 \dots\dots\dots\dots (1),$$
$$p\alpha^2 + q\beta^2 + 2r'\alpha\beta = 0 \dots\dots\dots\dots (2).$$

The polar reciprocal of (1) with respect to (2) is

$$\begin{vmatrix} 0, & p\alpha + r'\beta, & r'\alpha + q\beta \\ p\alpha + r'\beta, & u, & w' \\ r'\alpha + q\beta, & w', & v \end{vmatrix} = 0,$$

ANHARMONIC PROPERTIES. 117

or $u(r'\alpha + q\beta)^2 + v(p\alpha + v'\beta)^2 - 2w'(r'\alpha + q\beta)(p\alpha + r'\beta) = 0$.

Suppose that
$$u\alpha^2 + v\beta^2 + 2w'\alpha\beta = u(\alpha + k_1\beta)(\alpha + k_2\beta)$$
identically, i.e. that
$$v = uk_1k_2,$$
$$2w' = u(k_1 + k_2).$$

At the point of intersection of the line $\alpha + k_1\beta = 0$, with $\gamma = 0$, we have
$$\frac{\alpha}{k_1} = -\beta, \ \gamma = 0.$$

Taking the polar of this with respect to the curve (2) we get
$$(pk_1 - r')\alpha + (r'k_1 - q)\beta = 0.$$

If this be identical with $\alpha + k_2\beta = 0$, we get
$$pk_1 - r' = \frac{r'k_1 - q}{k_2},$$
or $pk_1k_2 - 2v'(k_1 + k_2) + q = 0$;
$$\therefore pv - 2r'w' + qu = 0,$$
the required condition.

The symmetry of this equation shews that (2) is also its own polar reciprocal with respect to (1), as ought to be the case.

19. If the point of intersection of the four straight lines do not coincide with one of the angular points of the triangle of reference, we have then only to express the condition that the range formed by their intersection with any one of its sides, $\gamma = 0$, for instance, be an harmonic range. If this be the case, the pencil formed by joining these four points with C will be an harmonic pencil, and we shall have, as before,
$$pv - 2r'w' + qu = 0.$$

20. We next proceed to consider the results to be deduced from the theory of reciprocal polars, when the auxiliary conic is a circle. It is here that the utility of theory is most apparent, as we are thus enabled to transform *metrical* theorems, i.e. theorems relating to the *magnitudes* of lines and angles.

We know that, if PQ be the polar of a point T with respect to a circle, of which the centre is S and radius k, then ST will be perpendicular to PQ. Let ST cut PQ in V. Then
$$ST \cdot SV = k^2.$$

Hence *the pole of any line is at a distance from the centre of the auxiliary circle inversely proportional to the distance of the line.* And conversely, *the polar of any point is at a distance from the centre of the auxiliary circle, inversely proportional to the distance of the point itself.*

21. If TX, TY be any two indefinite straight lines, P, Q their poles, then, since SP is perpendicular to TX, SQ to TY, it follows that the angle PSQ is equal to the angle XTY or its supplement, as the case may be. Hence, *the angle included between any two straight lines is equal to the angle subtended at the centre of the auxiliary circle by the straight line joining their poles, or to its supplement.*

22. From what has been said in Art. 15, and the earlier articles of this chapter, it will appear that to find the polar reciprocal of a given curve with respect to a circle, we may proceed by either of the following two methods.

First. Draw a tangent to the curve, and from S, the centre of the auxiliary circle, draw SY perpendicular to the tangent, and on SY, produced if necessary, take a point Q, such that $SQ \cdot SY = k^2$. The locus of Q will be the required polar reciprocal.

Secondly. Take a point P on the curve, and join SP; on SP, produced if necessary, take a point Z, such that
$$SP \cdot SZ = k^2.$$
Through Z draw a straight line perpendicular to SP. The envelope of this line will be the required polar reciprocal.

23. It will be observed that the magnitude of the radius of the auxiliary circle affects the absolute, but not the relative, magnitudes or positions of the various lines in the reciprocal figure. As our theorems are, for the most part, independent of absolute magnitude, we may generally drop all consideration of the radius of the auxiliary circle, and consider its centre only. We may then speak of reciprocating "with respect to S" instead of "with respect to a circle of which S is the centre." S may be called the *centre of reciprocation*, k the *constant of reciprocation*.

24. As an example of the power of this method we will reciprocate the following theorem, "The three perpendiculars from the angular points of a triangle intersect in a point."

This may be expressed as follows: "If O, A, B, C be four points, such that OB is perpendicular to CA, and OC to AB, then will OA be perpendicular to BC."

Reciprocate this with respect to any point S, and the four points O, A, B, C give four straight lines, which we may call each by three letters $abc, ab'c', a'bc', a'b'c$, respectively. Then, the fact that OB is perpendicular to CA is expressed by b and b' subtending a right angle at S, or by bSb' being a right angle. Again, the fact that OC is perpendicular to AB, shews that cSc' is a right angle. Then the reciprocal theorem tells us that aSa' is also a right angle. We may express this more neatly as follows: aa', bb', cc', are the diagonals of the complete quadrilateral formed by the four straight lines, hence it appears that at any point at which two of the diagonals of a complete quadrilateral subtend a right angle, the third diagonal also subtends a right angle. Or, in other words, *The three circles, described on the diagonals of a complete quadrilateral as diameters, have a common radical axis.*

The extremities of this axis may be conveniently called the *foci* of the quadrilateral[*].

25. If the system formed by the four points O, A, B, C be reciprocated with respect to any one of them, O for instance, the triangle thus obtained will be similar, and similarly situated, to that formed by the other three points A, B, C.

---

[*] This name is proposed by Clifford, in the *Messenger of Mathematics*.

For if on OA, OB, OC respectively (produced if necessary), we take points A', B', C', so that
$$OA \cdot OA' = OB \cdot OB' = OC \cdot OC',$$
and through A', B', C' draw YZ, ZX, XY, severally at right angles to OA', OB', OC', then YZ, ZX, XY are respectively parallel to BC, CA, AB, or the triangle XYZ is similar and similarly situated to the triangle ABC.

We may observe further, that the point X, since it is the intersection of the polars of B and C, is itself the pole of the line BC, and therefore OX is perpendicular to BC, that is to YZ. Similarly, OY, OZ, are respectively perpendicular to ZX, XY. Hence, O is the intersection of the perpendiculars dropped from X, Y, Z on YZ, ZX, XY respectively. It may be convenient to call the point of intersection of the perpendiculars let fall from the angular points of a triangle on the opposite sides, the *orthocentre* of the triangle, or of its three angular points. Here we may say that "If a triangle be reciprocated with respect to its orthocentre, the reciprocal triangle will be similar and similarly situated to the given triangle, and will have the same orthocentre."

It will be seen by Art. 19, that any three points and their orthocentre, reciprocated with respect to any point S, give a quadrilateral, of which S is a focus.

26. If any conic be reciprocated with respect to an external point S, the angle between the asymptotes of the reciprocal hyperbola will be the supplement of that between the tangents drawn from S to the conic. (See Art. 9 of this chapter.)

Conversely, if an hyperbola be reciprocated with respect to any point S, we obtain a conic, which subtends at S an angle the supplement of that between the asymptotes of the hyperbola.

27. From the last article it follows that, if a parabola be reciprocated with respect to any point S on its directrix, we obtain a rectangular hyperbola, passing through S.

If a rectangular hyperbola be reciprocated with respect to

RECIPROCATION WITH RESPECT TO A CIRCLE.   121

any point S on its circumference, we obtain a parabola whose directrix passes through S.

Again, if a conic be reciprocated with respect to any point on its director circle (i. e. the circle which is the locus of the intersection of two perpendicular tangents) we obtain a rectangular hyperbola.

If a rectangular hyperbola be reciprocated with respect to any point S not on the curve, we obtain a conic, whose director circle passes through S.

28. It is known that the conics passing through the four points of intersection of any two rectangular hyperbolas, is itself a rectangular hyperbola; and also that any one of these four points is the orthocentre of the other three. If, then, we reciprocate these theorems with respect to any one of the four points of intersection, we obtain the theorem that, "If a parabola touch the three common tangents of two given parabolas, its directrix passes through the intersection of the directrices of the two given parabolas, that is, through the orthocentre of the triangle formed by their common tangents." In other words, "If a system of parabolas be described, touching three given straight lines, their directrices all pass through the orthocentre of the triangle formed by the three given straight lines."

Again, reciprocating this system of rectangular hyperbolas with respect to any point S, we get, "All conics, which touch four given straight lines, subtend a right angle at either focus of the quadrilateral formed by these four straight lines." Or, in other words, "The director circles of all conics which touch four given straight lines, have a common radical axis, which is the directrix of the parabola which touches the four given straight lines."

29. *To find the polar reciprocal of a circle with respect to any point.*

From what has already been shewn, we know that this will be a conic; we have now to investigate its form and position.

Let S be the centre of reciprocation, k the constant of reciprocation, MPM' the circle to be reciprocated, O its centre,

MM' its diameter passing through S, ρ its radius, and let $OS = c$.

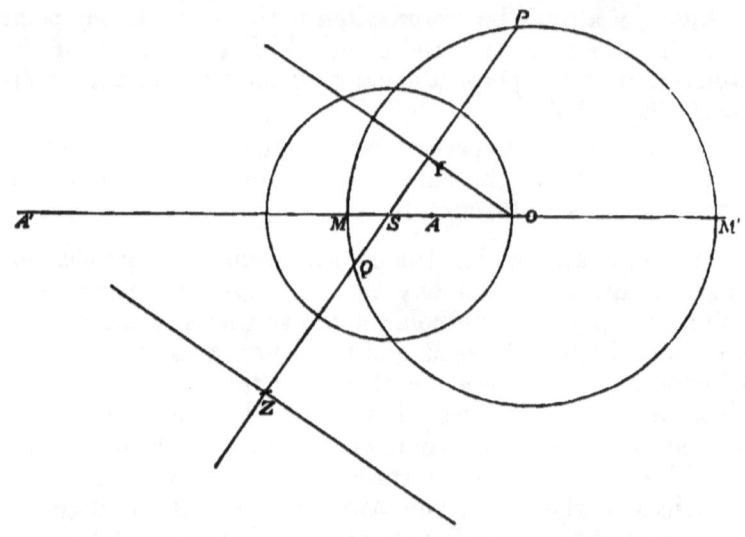

Fig. 19.

Through S draw any straight line cutting MPM' in P and Q.

On SPQ, produced if necessary, take two points Y and Z, such that
$$SP . SY = SQ . SZ = k^2.$$
The straight lines drawn through Y and Z perpendicular to SP will be tangents to the reciprocal conic.

Now $$SY . SZ = \frac{k^4}{SP . SQ} = \frac{k^4}{\rho^2 - c^2},$$
which is constant. Hence, the reciprocal is a conic of such a nature that the rectangle under the distances from S of any two parellel tangents is constant. It is therefore a conic, of which S is a focus, and of which the axis-minor is $\dfrac{2k^2}{(\rho^2 - c^2)^{\frac{1}{2}}}$.

It will be an ellipse, parabola, or hyperbola, according as ρ is greater than, equal to, or less than c, that is, according as

RECIPROCATION OF A CIRCLE WITH RESPECT TO A POINT. 123

the centre of reciprocation lies within, upon, or without, the circle to be reciprocated. This agrees with what has been already shewn, Art. 9.

Let $2a$, $2b$, be the axes of the conic, $2l$ its latus-rectum, e its eccentricity.

To determine their magnitudes, we proceed as follows. The axis-major will be in the direction SO. Let A, A' be its extremities.

Then $\dfrac{2}{l} = \dfrac{1}{SA} + \dfrac{1}{SA'} = \dfrac{SM + SM'}{k^2} = \dfrac{2\rho}{k^2}$.

Hence, $l = \dfrac{k^2}{\rho}$, or the latus-rectum is inversely proportional to the radius of the circle.

Again, $a = \dfrac{b^2}{l} = \dfrac{k^2 \rho}{\rho^2 - c^2}$,

$$e^2 = 1 - \dfrac{b^2}{a^2}$$

$$= 1 - \dfrac{k^4}{\rho^2 - c^2} \cdot \dfrac{(\rho^2 - c^2)^2}{k^4 \rho^2}$$

$$= \dfrac{c^2}{\rho^2},$$

or $e = \dfrac{c}{\rho}$.

Thus the eccentricity varies directly as the distance of the centre of the circle from the centre of reciprocation, and inversely as the radius of the circle.

If d be the distance from S of the corresponding directrix,

$$d = \dfrac{l}{e} = \dfrac{k^2}{\rho} \cdot \dfrac{\rho}{c} = \dfrac{k^2}{c},$$

or, the directrix is the polar of the centre of the circle.

30. We have now the means of obtaining, from any property of a circle, a focal property of a conic section.

Take, for example, Euc. III. 21. This may be expressed as follows: "If three points be taken on the circumference of a circle, two fixed and the third moveable, the straight lines joining the moveable point with the two fixed points, make a constant angle with one another." This will be reciprocated into "If three tangents be drawn to a conic section, two fixed and the third moveable, the portion of the moveable tangent intercepted between the two fixed ones, subtends a constant angle at the focus." This angle will be found, by reciprocating Euc. III. 20, to be the complement of one-half of the angle subtended at the focus by the portion of the corresponding directrix intercepted between the two fixed tangents.

'Again, it is easy to see that "if a circle be described touching two concentric circles, its radius will be equal to half the sum, or half the difference, of the radii of the given circles, and the locus of its centre will be a circle, concentric with the other two, and of which the radius is half the difference, or half the sum, of the radii of the two given circles."

Hence we deduce the following theorem. "If two conics have a common focus and directrix, and their latera-recta be $2l$, $2l'$, and another conic, having the same focus, be described so as to touch both of them, its latus-rectum will be $\dfrac{4ll'}{l \mp l'}$, and the envelope of its directrix will be a conic, having the same focus and directrix as the given conics, and of which the latus-rectum is $\dfrac{4ll'}{l \mp l'}$."

Again, take the ordinary definition of an ellipse, that it is the locus of a point, the sum of the distances of which from two fixed points is constant. This is equivalent to "the sum of the distances from either focus, of the points of contact of two parallel tangents, is constant."

The reciprocal theorem will be, "If a system of chords be drawn to a circle, passing through a given point, and, at the extremities of any chord, a pair of tangents be drawn to the circle, the sum of the reciprocals of the distances of these tangents from the fixed point is constant."

FOCAL PROPERTIES. 125

The known property of a circle, that "two tangents make equal angles with their chord of contact," will be found, when transformed by the method now explained, to be equivalent to the theorem that "if two tangents be drawn to a conic from an external point, the portions of these tangents, intercepted between that point and their points of contact, subtend equal angles at the focus." From the fact that "all circles intersect in two imaginary points at infinity," we learn that "all conics, having a common focus, have a common pair of imaginary tangents passing through that focus." And, more generally, we may say that all similar and similarly situated conics reciprocate into a system of conics having two common tangents.

31. Two points, on a curve and its reciprocal, are said to *correspond* to one another when the tangent at either point is the polar of the other point. Two tangents are said to correspond when the point of contact of either is the pole of the other.

The angle between the radius vector of any point (drawn from the centre of reciprocation), and the tangent at that

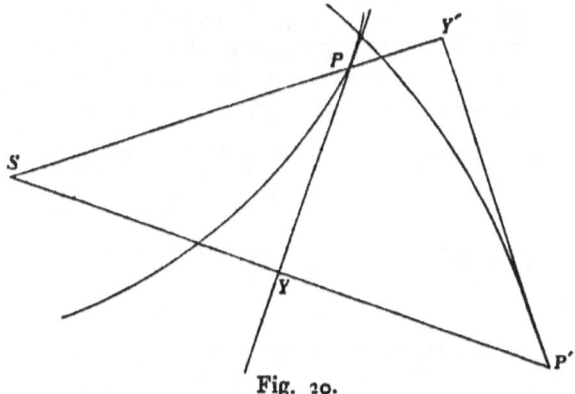

Fig. 20.

point, is equal to the angle between the radius vector of, and tangent at, the corresponding point of the reciprocal curve.

For, if P be the given point, PY the tangent at P, and S the centre of reciprocation, and SY be perpendicular to PY; and if P' be the pole of PY, and $P'Y'$ the polar of P, then

126   TRILINEAR CO-ORDINATES.

P' will lie on SY, produced if necessary; and if SY' be perpendicular to $P'Y'$, SY' will pass through P. Hence, since SP, PY, are respectively perpendicular to $P'Y'$, SP', it follows that the angle SPY is equal to the angle $SP'Y'$.

32. We have investigated (Art. 10, Chap. IV.) the equation of the two tangents drawn to a conic from any given point (f, g, h). If in the right-hand member of that equation we substitute for 0, $\omega\,(a\alpha + b\beta + c\gamma)^2$, ω being an arbitrary constant, we shall obtain the general equation of all conics of which these lines are asymptotes. Now, since the asymptotes of the reciprocal conic with respect to (f, g, h), are respectively at right angles to the two tangents drawn from (f, g, h), it follows that the family of conics thus obtained will be similar in form to the reciprocal conic.

33. *To find the co-ordinates of the foci of the conic represented by the general equation of the second degree.*

Since the reciprocal of a conic with respect to a focus is a circle, it will follow from Art. 32 that the family of conics obtained as above must, if (f, g, h) be a focus, be circles also. Applying the conditions for a circle investigated in Art. 14, Chap. IV., it will be found that the terms involving ω disappear of themselves, and our conditions assume the form

$$(Uh^2 + Wf^2 - 2V'hf)c^2 + (Vf^2 + Ug^2 - 2W'fg)b^2$$
$$+ 2(U'f^2 + Ugh - W'hf - V'fg)bc$$
$$= (Vf^2 + Ug^2 - 2W'fg)a^2 + (Wg^2 + Vh^2 - 2U'gh)c^2$$
$$+ 2(V'g^2 + Vhf - U'fg - W'gh)ca$$
$$= (Wg^2 + Vh^2 - 2U'gh)b^2 + (Uh^2 + Wf^2 - 2V'hf)a^2$$
$$+ 2(W'h^2 + Wfg - V'gh - U'hf)ab,$$

or

$$(Vb^2 + Wc^2 + 2U'bc)f^2 - 2(V'c + W'b)f(bg + ch) + U(bg + ch)^2$$
$$= (Wc^2 + Ua^2 + 2V'ca)g^2 - 2(W'a + U'c)g(ch + af) + V(ch + af)^2$$
$$= (Ua^2 + Vb^2 + 2W'ab)h^2 - 2(U'b + V'a)h(af + bg) + W(af + bg)^2,$$

FOCI OF A CONIC.

equations which, since $af + bg + ch = 2\Delta$, may also be written under the form

$$(Ua^2 + Vb^2 + Wc^2 + 2U'bc + 2V'ca + 2W'ab)f^2$$
$$- 4\Delta(V'c + W'b + Ua)f + 4U.\Delta^2$$
$$= (Ua^2 + Vb^2 + Wc^2 + 2U'bc + 2V'ca + 2W'ab)g^2$$
$$- 4\Delta(W'a + U'c + Vb)g + 4V.\Delta^2$$
$$= (Ua^2 + Vb^2 + Wc^2 + 2U'bc + 2V'ca + 2W'ab)h^2$$
$$- 4\Delta(U'b + Va + Wc)h + 4W.\Delta^2.$$

The equations, together with

$$af + bg + ch = 2\Delta,$$

determine the co-ordinates of the foci. It will be seen that they give four values of f, g, h, two of which are real, two imaginary.

If the conic be a parabola, then, applying the condition of Art. 6, Chap. IV., these equations reduce to

$$(V'c + W'b + Ua)f - U\Delta = (W'a + U'c + Vb)g - V\Delta$$
$$= (U'b + V'a + Wc)h - W\Delta,$$

which give the focus in that case.

If the equation

$$ux^2 + vy^2 + wz^2 + 2u'yz + 2v'zx + 2w'xy = 0,$$

be expressed in triangular co-ordinates, we get, for the co-ordinates of the foci, the equations

$$\frac{(U + V + W + 2U' + 2V' + 2W')f^2 - 2(V' + W' + U)f + U}{a^2}$$
$$= \frac{(U + V + W + 2U' + 2V' + 2W')g^2 - 2(W' + U' + V)g + V}{b^2}$$
$$= \frac{(U + V + W + 2U' + 2V' + 2W')h^2 - 2(U' + V' + W)h + W}{c^2},$$

or, if the conic be a parabola,

$$\frac{2(V'+W'+U)f-U}{a^2} = \frac{2(W'+U'+V)g-V}{b^2}$$
$$= \frac{2(U'+V'+W)h-W}{c^2}.$$

34. Interesting results may sometimes be obtained by a double application of the method of reciprocal polars. Thus, the theorem that "the angle in a semicircle is a right angle" may be expressed in the form that "every chord of a circle, which subtends a right angle at a given point of the curve, passes through the centre." Reciprocating this with respect to the given point, we get

"The locus of the point of intersection of two tangents to a parabola at right angles to one another, is the directrix."

Now, reciprocate this with respect to *any point whatever*, and we find that

"Every chord of a conic which subtends a right angle at a given point on the curve, passes through a fixed point."

Again, take Euc. III. 21. This may be expressed under the form "If a chord be drawn to a circle subtending a constant angle at a fixed point O on its circumference, it always touches a concentric circle." Reciprocating this theorem with respect to O, we get "If two tangents be drawn to a parabola containing a constant angle, the locus of their point of intersection will be a conic, having a focus and directrix in common with the given parabola." Reciprocate this, with respect to *any point whatever*, and we get, "If a chord be drawn to a conic, subtending a constant angle at a given point on the curve, it always touches a conic having double contact with the given one."

EXAMPLES.

1. Having a given focus and two points of a conic section, prove that the locus of the point of intersection of the tangents at these points will be two straight lines, passing through the focus, and at right angles to each other.

EXAMPLES.

2. Prove that four conics can be described with a given focus and passing through three given points, and that the latus-rectum of one of these is equal to the sum of the latera-recta of the other three.

3. On a fixed tangent to a conic are taken a fixed point A, and two moveable points P, Q, such that AP, AQ subtend equal angles at a fixed point O. From P, Q are drawn two other tangents to the conic, prove that the locus of their point of intersection is a straight line.

4. Two variable tangents are drawn to a conic section so that the portion of a fixed tangent, intercepted between them, subtends a right angle at a fixed point. Prove that the locus of the point of intersection of the variable tangents is a straight line.

If the fixed point be a focus, the locus will be the corresponding directrix.

5. Chords are drawn to a conic, subtending a right angle at a fixed point; prove that they all touch a conic, of which that point is a focus.

6. Three given straight lines BC, CA, AB are intersected by two other given straight lines in A_1, A_2; B_1, B_2; C_1, C_2 respectively. Prove that a conic can be described touching the six straight lines AA_1, AA_2, BB_1, BB_2, CC_1, CC_2.

7. A, B, C, S are four fixed points, SD is drawn perpendicular to SA, intersecting BC in D, SE perpendicular to SB, intersecting CA in E, SF perpendicular to SC, intersecting AB in F. Prove that D, E, F lie in the same straight line.

Prove also that the four conics which have S as a focus, and which touch the three sides of the several triangles ABC, AEF, BFD, CDE have their latera-recta equal.

8. Two conics are described with a common focus and their corresponding directrices fixed; prove that, if the sum of the reciprocals of their latera-recta be constant, their common tangents will touch a conic section.

9. A conic is described touching three given straight lines BC, CA, AB, so that the pair of tangents drawn to it from a given point O, are at right angles to each other. Prove that it will always touch another fixed straight line; and that, if this straight

line cut BC, CA, AB in D, E, F respectively, each of the angles AOD, BOE, COF is a right angle.

Prove also that the polar of O with respect to this conic will always touch a conic, of which O is a focus.

10. OA, OB are the common tangents to two conics having a common focus S, CA, CB are tangents at one of their points of intersection, BD, AE tangents intersecting CA, CB in D, E. Prove that S, D, E lie in the same straight line.

11. Any triangle is described, self-conjugate with regard to a given conic; prove that, if a conic be described, touching the sides of this triangle, and having the centre of the given conic as a focus, its axis-minor will be constant.

12. Prove that two ellipses, which have a common focus, cannot intersect in more than two points.

13. If a system of conics be described, passing through four given points, four fixed straight lines may be found, such that the chord of each, intercepted by any conic of the system, subtends a right angle at one of the points.

14. If a parabola be reciprocated with respect to any circle, the radius of curvature of the reciprocal conic, at the origin, varies inversely as the latus-rectum of the parabola.

15. If the two pairs of straight lines represented by the equations
$$u\alpha^2 + v\beta^2 + 2w'\alpha\beta = 0,$$
$$p\alpha^2 + q\beta^2 + 2r'\alpha\beta = 0,$$
form a harmonic pencil, the two straight lines of each pair not being conjugate, prove that
$$(uq + vp - 2w'r')^2 = 36\,(uv - w'^2)(pq - r'^2).$$

CHAPTER VII.

TANGENTIAL CO-ORDINATES.

1. In the systems of co-ordinates with which we have hitherto been concerned, we have considered a point as determined, directly or indirectly, by means of its distances from three given straight lines; and we have regarded a curve as the aggregation of all points, the co-ordinates of which satisfy a certain equation. It is equally possible, however, to consider a straight line as determined by means of its distances from three points, which distances may be termed its co-ordinates; and to regard a curve as the envelope of all straight lines, the co-ordinates of which satisfy a certain equation.

This system is closely connected with the theory of reciprocal polars. In fact, it may be looked upon as a means of so interpreting equations as at once to obtain the results which the method of reciprocal polars would deduce from the ordinary method of interpretation. The equations are the reciprocals of those described in Chapter v. with respect to $x^2 + y^2 + z^2 = 0$.

We may then define the co-ordinates of a straight line to be the perpendiculars let fall upon it from three given points A, B, C. The lengths of these perpendiculars we will denote by the letters p, q, r respectively, the lengths BC, CA, AB being represented as before by the letters a, b, c, and the angles of the triangle of reference ABC being denoted by A, B, C, and its area by Δ.

2. Any two co-ordinates, q and r for example, will be considered to have contrary signs if the line of which they

are the co-ordinates cuts the line BC in a point lying between B and C, otherwise to have the same sign. Thus, the internal bisector of the angle A has its co-ordinates of contrary signs, the external bisector of the same sign. The sign of p relatively to q and r will be determined in the same manner.

If D be any point on the line BC, q, r the co-ordinates of any line passing through it, and $BD = a_1$, $CD = a_2$, distances measured along the line BC from B to C being considered positive, and from C to B negative, it will readily be seen that

$$\frac{q}{a_1} = \frac{r}{a_2}.$$

Since this is a relation between the co-ordinates of any line passing through the point D, it may be considered as the *equation* of the point D.

If D be the middle point of BC, $a_1 = -a_2$, hence it appears that the middle points of the sides of the triangle of reference are represented by the equations,

$$q + r = 0, \quad r + p = 0, \quad p + q = 0.$$

It may also be proved that the points where the internal bisectors of the angles meet the opposite sides, are represented by

$$bq + cr = 0, \quad cr + ap = 0, \quad ap + bq = 0.$$

The points where the external bisectors of the angles meet the opposite sides, by

$$bq - cr = 0, \quad cr - ap = 0, \quad ap - bq = 0.$$

The feet of the perpendiculars from the angular points on the opposite sides, by

$$q \tan B + r \tan C = 0, \quad r \tan C + p \tan A = 0,$$
$$p \tan A + q \tan B = 0.$$

The points of contact of the inscribed circle, by

$$\frac{q}{s-b} + \frac{r}{s-c} = 0, \quad \frac{r}{s-c} + \frac{p}{s-a} = 0, \quad \frac{p}{s-a} + \frac{q}{s-b} = 0,$$

where $2s = a + b + c$.

EQUATION OF A POINT.

3. We shall next prove the following proposition; that if O be any given point within the triangle ABC, then the co-ordinates p, q, r (their signs being taken in the manner already explained) of any straight line QPR, passing through it, will be connected by the following equation,

$$\triangle BOC . p + \triangle COA . q + \triangle AOB . r = 0.$$

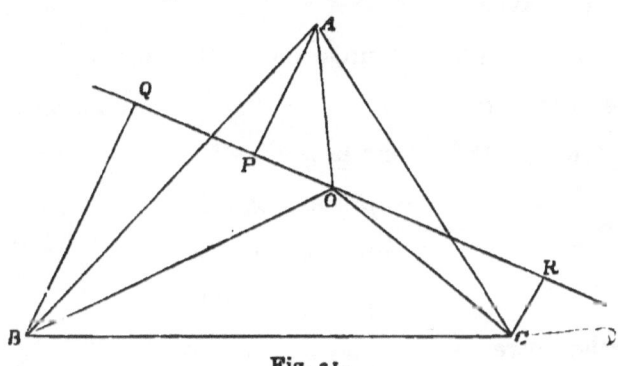

Fig. 21.

Let the triangular equation of QPR be

$$lx + my + nz = 0,$$

and the trilinear co-ordinates of O, ξ, η, ζ.

Then $\xi : \eta : \zeta :: \triangle BOC : \triangle COA : \triangle AOB$.

And, since O lies on QPR,

$$l\xi + m\eta + n\zeta = 0.$$

Again, since p is the distance from the point $(1, 0, 0)$ to the line, (l, m, n),

$$\therefore (l.2\Delta)^2 = \{(l-m)(l-n)a^2 + (m-n)(m-l)b^2 + (n-l)(n-m)c^2\}p^2.$$

Similar equations hold for Q and R, hence

$$\frac{p}{l} = \frac{q}{m} = \frac{r}{n},$$

or $p\xi + m\eta + n\zeta = 0$,

whence $\quad \triangle BOC . p + \triangle COA . q + \triangle AOB . r = 0.$

This equation may be regarded as the equation of the point O.

A similar equation may be proved to hold for any point without the triangle, BOC being considered negative, if A and O be on opposite sides of BC.

The following are the equations of some important points connected with the triangle of reference:

Centre of gravity, $\hspace{4em} p + q + r = 0.$

Centre of circumscribing circle, $p\sin 2A + q\sin 2B + r\sin 2C = 0.$

Centre of inscribed circle, $\hspace{4em} ap + bq + cr = 0.$

Centres of escribed circles,
$$-ap + bq + cr = 0,$$
$$ap - bq + cr = 0,$$
$$ap + bq - cr = 0.$$

Orthocentre,
$$p\tan A + q\tan B + r\tan C = 0.$$

4. We proceed to investigate the identical relation which holds between the co-ordinates of any straight line.

Let any straight line cut the sides AB, AC of the triangle of reference in D, E. From A, B, C let fall AP, BQ, CR, perpendiculars on the line, then $BQ = q$, $CR = r$, $AP = -p$.

Let the triangular equation of RPQ referred to ABC as triangle of reference, be
$$lx + my + nz = 0.$$

Then, as shewn in the last article,
$$(l.2\Delta)^2 = \{(l-m)(l-n)a^2 + (m-n)(m-l)b^2 + (n-l)(n-m)c^2\}p^2.$$

Similar equations holding for q^2 and r^2, we get
$$\frac{p^2}{l^2} = \frac{q^2}{m^2} = \frac{r^2}{n^2}$$
$$= \frac{(2\Delta)^2}{(l-m)(l-n)a^2 + (m-n)(m-l)b^2 + (n-l)(n-m)c^2},$$

IDENTICAL RELATION.

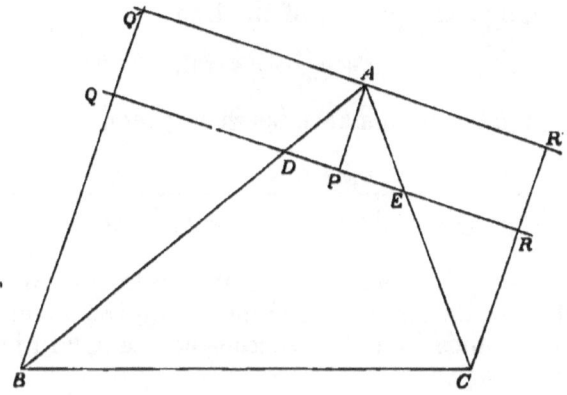

Fig. 22.

whence

$$(p-q)(p-r)a^2 + (q-r)(q-p)b^2 + (r-p)(r-q)c^2 = 4\Delta^2,$$

the required relation between the co-ordinates of any straight line whatever.

This may also be expressed in the form

$$\begin{vmatrix} 0, & 1, & 1, & 1, & 0 \\ 1, & 0, & c^2, & b^2, & p \\ 1, & c^2, & 0, & a^2, & q \\ 1, & b^2, & a^2, & 0, & r \\ 0, & p, & q, & r, & 1 \end{vmatrix} = 0,$$

as may be found by evaluating the determinant.

COR. Since the line at infinity may be considered as equidistant from A, B, and C, it will be represented by the equations $p = q = r$.

5. *To find the distance from the point* $lp + mq + nr = 0$, *to the line* (p_1, q_1, r_1).

By what has been shewn above, it appears that the triangular co-ordinates of the point are

$$\frac{l}{l+m+n}, \quad \frac{m}{l+m+n}, \quad \frac{n}{l+m+n}.$$

And the triangular equation of the line

$$p_1 x + q_1 y + r_1 z = 0.$$

Hence, if δ be the distance between them,

$$\delta = \frac{p_1 l + q_1 m + r_1 n}{(l+m+n)\{(p_1-q_1)(p_1-r_1)a^2+(q_1-r_1)(q_1-p_1)b^2+(r_1-p_1)(r_1-q_1)c^2\}^{\frac{1}{2}}} 2\Delta.$$

Cor. Hence, if ρ be the radius of a circle, $lp + mq + nr = 0$ the equation of its centre, the circle, being the envelope of a line whose distance from the centre is constant, will be represented by the equation

$$(p-q)(p-r)a^2 + (q-r)(q-p)b^2 + (r-p)(r-q)c^2$$
$$= \left(\frac{2\Delta}{\rho}\right)^2 \left(\frac{lp+mq+nr}{l+m+n}\right)^2.$$

6. An equation of a degree, higher than the first, may be regarded as representing the curve which is touched by all the straight lines, the co-ordinates of which satisfy the equation of the curve. Adopting this mode of interpretation, the values of the ratios $p : q : r$ which simultaneously satisfy two given equations will be the co-ordinates of the common tangents to the two curves represented by these equations, and the values obtained by combining any given equation with an equation of the first degree, will represent all the straight lines which pass through the point represented by the equation of the first degree, and which touch the curve. From this it follows, that an equation of the nth degree will represent a curve such that n tangents, real or imaginary, can be drawn to it from any point, that is, a curve of the nth class.

It will hence follow that every equation of the second degree represents a conic. We may proceed to consider some of its more interesting special forms.

7. *To find the equation of a conic which touches the three sides of the triangle of reference.*

EQUATION OF THE SECOND DEGREE.

The co-ordinates of the sides of the triangle of reference are

$$q = 0, \quad r = 0 \text{ for } BC,$$
$$r = 0, \quad p = 0 \text{ for } CA,$$
$$p = 0, \quad q = 0 \text{ for } AB.$$

Hence, the equation of the required conic must be satisfied whenever two out of the three co-ordinates p, q, r are $= 0$. It must therefore be of the form

$$Lqr + Mrp + Npq = 0.$$

The equations of the points of contact are

$$\frac{q}{M} + \frac{r}{N} = 0,$$

$$\frac{r}{N} + \frac{p}{L} = 0,$$

$$\frac{p}{L} + \frac{q}{M} = 0.$$

These may be established as follows: If in the given equation we make $Mr + Nq = 0$, we obtain either $q = 0$, or $r = 0$. It hence appears that the tangents drawn through the point $Mr + Nq = 0$, pass either through the point $q = 0$, or through the point $r = 0$. But the three points

$$Mr + Nq = 0, \quad q = 0, \quad r = 0,$$

lie in the same straight line; hence the tangents drawn from $Mr + Nq = 0$ coincide, that is, it is the point of contact of the tangent for which $q = r = 0$. Similarly for the other two points of contact.

It will hence appear, by reference to the equations of the points of contact of the inscribed circle, given in Art. 2, that that circle is represented by the equation

$$(s-a)\,qr + (s-b)\,rp + (s-c)\,pq = 0.$$

The escribed circles will be represented as follows:

$$-sqr + (s-c)rp + (s-b)pq = 0,$$
$$(s-c)qr - srp + (s-a)pq = 0,$$
$$(s-b)qr + (s-a)rp - spq = 0.$$

8. *To find the equation of a conic circumscribed about the triangle of reference.*

The equations of the angular points of the triangle of reference are $p = 0$, $q = 0$, $r = 0$. Now, since each of these points lies on the curve, the two tangents drawn through any one of them must coincide, hence when any one of these quantities is put $= 0$, the remaining equation must have two equal roots. The required equation will therefore be of the form

$$L^2 p^2 + M^2 q^2 + N^2 r^2 - 2MNqr - 2NLrp - 2LMpq = 0.$$

The co-ordinates of the several tangents at the angular points will be given by the equations

$$p = 0, \quad Mq - Nr = 0,$$
$$q = 0, \quad Nr - Lp = 0,$$
$$r = 0, \quad Lp - Mq = 0.$$

If the conic be a circle, the tangent at A will be determined by the equations

$$p = 0, \quad \frac{q}{c \sin C} = \frac{r}{b \sin B},$$

which last is equivalent to $b^2 q - c^2 r = 0$.

Similar equations holding for the other two tangents, the equation for the circumscribing circle will be

$$a^4 p^2 + b^4 q^2 + c^4 r^2 - 2b^2 c^2 qr - 2c^2 a^2 rp - 2a^2 b^2 pq = 0,$$

which may be reduced to

$$\pm ap^{\frac{1}{2}} \pm bq^{\frac{1}{2}} \pm cr^{\frac{1}{2}} = 0.$$

9. By investigations similar to those in Chap. IV. Art. 8, it may be shewn that the equation of the pole of the line (f, g, h) with respect to the conic

$$\phi(p, q, r) = up^2 + vq^2 + wr^2 + 2u'qr + 2v'rp + 2w'pq = 0,$$

is

$$(uf + w'g + v'h)p + (w'f + vg + u'h)q + (v'f + u'g + wh)r = 0.$$

Now, the centre is the pole of the line at infinity, which is given by the equations $p = q = r$.

The equation of the centre is therefore

$$(u + v' + w')p + (u' + v + w')q + (u' + v' + w)r = 0.$$

If the conic be a parabola, it touches the line at infinity; the condition that it should be a parabola is therefore

$$u + v + w + 2u' + 2v' + 2w' = 0.$$

10. The two points in which the conic is cut by the line (f, g, h) are represented by the equation

$$4\phi(f, g, h)\phi(p, q, r) - \left(p\frac{d\phi}{df} + q\frac{d\phi}{dg} + r\frac{d\phi}{dh}\right)^2 = 0.$$

(See Chap. IV. Art. 10.)

Hence the two points in which it is cut by the line at infinity are given by

$$(u+v+w+2u'+2v'+2w')(up^2+vq^2+wr^2+2u'qr+2v'rp+2w'pq)$$
$$- \{(u + v' + w')p + (u' + v + w')q + (u' + v' + w)r\}^2 = 0.$$

Hence may be deduced the equation of the two points at infinity through which all circles pass. For these are the same for all circles. Now, for the inscribed circle they are obtained by putting

$$u = v = w = 0, \quad 2u' = s - a, \quad 2v' = s - b, \quad 2w' = s - c.$$

The equation then becomes

$$4s\{(s - a)qr + (s - b)rp + (s - c)pq\} - (ap + bq + cr)^2 = 0,$$

or $a^2p^2 + b^2q^2 + c^2r^2 - 2bcqr \cos A - 2carp \cos B - 2abpq \cos C = 0,$

which may also be written
$$a^2(p-q)(p-r) + b^2(q-r)(q-p) + c^2(r-p)(r-q) = 0,$$
the equation of the two circular points at infinity.

Hence may also be deduced the conditions that the equation
$$up^2 + vq^2 + wr^2 + 2u'qr + 2v'rp + 2w'pq = 0$$
may represent a circle. For, comparing this equation with that just obtained (Art. 5, Cor.), we get
$$\frac{a^2 - l^2}{u} = \frac{b^2 - m^2}{v} = \frac{c^2 - n^2}{w}$$
$$= \frac{a^2 - b^2 - c^2 - 2mn}{2u'} = \frac{b^2 - c^2 - a^2 - 2nl}{2v'} = \frac{c^2 - a^2 - b^2 - 2lm}{2w'}.$$

Putting each member of these equations $= k$, they may be written
$$a^2 - uk = l^2, \quad b^2 - vk = m^2, \quad c^2 - wk = n^2,$$
$$bc \cos A + u'k = -mn, \quad ca \cos B + v'k = -nl,$$
$$ab \cos C + w'k = -lm.$$

These equations give
$$(b^2 - vk)(c^2 - wk) = (bc \cos A + u'k)^2,$$
or $(vw - u'^2)k^2 - (vc^2 + wb^2 + 2u'bc \cos A)k + b^2c^2 \sin^2 A = 0.$

Again, we get
$$(a^2 - uk)(bc \cos A + u'k) + (ca \cos B + v'k)(ab \cos C + w'k) = 0,$$
or $(v'w' - uu')k^2 + \{a(au' + b\cos C.v' + c\cos B.w') - bc\cos A.u\}k + a^2bc \sin B \sin C = 0.$

Now, since $b^2c^2 \sin^2 A = a^2bc \sin B \sin C = (2\Delta)^2$, these equations may be written under the form
$$(vw - u'^2)k^2 - (vc^2 + wb^2 + 2u'bc \cos A)k + 4\Delta^2 = 0,$$
$(v'w' - uu')k^2 + \{a(au' + bv'\cos C + cw'\cos B) - bc\cos A.u\}k + 4\Delta^2 = 0.$

CIRCULAR POINTS AT INFINITY.   141

Combining these with the four similar equations, we get

$$\{wu - v'^2 + uv - w'^2 - 2(v'w' - uu')\} k$$
$$- \{u(b^2 + c^2 - 2bc \cos A) + va^2 + wa^2 + 2a^2 u'$$
$$+ 2av'(c \cos B + b \cos C) + 2aw'(b \cos C + c \cos B)\} = 0,$$

or $\dfrac{wu - v'^2 + uv - w'^2 - 2(v'w' - uu')}{(u + v + w + 2u' + 2v' + 2w')a^2} = \dfrac{1}{k}.$

Two other corresponding expressions may of course be obtained for k, and the required condition is therefore

$$\frac{wu - v'^2 + uv - w'^2 - 2(v'w' - uu')}{a^2} = \frac{uv - w'^2 + vw - u'^2 - 2(w'u' - vv')}{b^2}$$
$$= \frac{vw - u'^2 + wu - v'^2 - 2(u'v' - ww')}{c^2}.$$

11. *To find the equation of the conic with respect to which the triangle of reference is self-conjugate.*

Since each angular point of the triangle of reference is the pole, with respect to this conic, of the opposite side, it follows that the equation of such a conic will be of the form

$$up^2 + vq^2 + wr^2 = 0.$$

From the last article it will be seen that the equation of the self-conjugate circle is

$$\frac{p^2}{b^2 + c^2 - a^2} + \frac{q^2}{c^2 + a^2 - b^2} + \frac{r^2}{a^2 + b^2 - c^2} = 0,$$

or $\dfrac{p^2}{bc \cos A} + \dfrac{q^2}{ca \cos B} + \dfrac{r^2}{ab \cos C} = 0,$

which may also be written

$$p^2 \tan A + q^2 \tan B + r^2 \tan C = 0.$$

EXAMPLES.

1. A parabola is described about a triangle so that the tangent at one angular point is parallel to the opposite side; shew that the square roots of the perpendiculars on any tangent to the curve are in arithmetical progression.

2. A conic is circumscribed about a triangle such that the tangent at each angular point is parallel to the opposite side; shew that, if p, q, r be the perpendiculars from the angular points on any tangent,
$$\pm p^{\frac{1}{2}} \pm q^{\frac{1}{2}} \pm r^{\frac{1}{2}} = 0.$$

3. Shew that the equation of the centre of this conic is
$$p + q + r = 0.$$

4. Conics are drawn each touching two sides of a triangle at the angular points and intersecting in a point; prove that the intersections of the tangents at this common point with the sides cutting their respective conics lie on one straight line, and that the common tangents to the conics intersect the sides in the same three points.

5. A system of hyperbolas is described about a given triangle; prove that, if one of the asymptotes always pass through a fixed point, the other will always touch a fixed conic, to which the three sides of the triangle are tangents.

6. A parabola touches one side of a triangle in its middle point, and the other two sides produced; prove that the perpendiculars, drawn from the angular points of the triangle upon any tangent to the parabola, are in harmonical progression.

7. Prove that the nine-point circle of the triangle of reference is represented by the equation
$$a\,(q+r)^{\frac{1}{2}} + b\,(r+p)^{\frac{1}{2}} + c\,(p+q)^{\frac{1}{2}} = 0.$$

8. If ρ_1, ρ_2, ρ_3 be the radii of three circles, the internal common tangents to each pair of the circles touch a conic whose tangential equation, referred to the centres of the circles, is
$$a^2\,(p-q)\,(p-r) + b^2\,(q-r)\,(q-p) + c^2\,(r-p)\,(r-q)$$
$$+ 4\Delta^2 \left(\frac{qr}{\rho_2 \rho_3} + \frac{rp}{\rho_3 \rho_1} + \frac{pq}{\rho_1 \rho_2} \right) = 0.$$

12. There is another system of Tangential Co-ordinates, which bears a close analogy to the ordinary Cartesian system. If x, y be the Cartesian co-ordinates of a point, referred to two rectangular axes, then the intercepts on these axes of the

TANGENTIAL RECTANGULAR CO-ORDINATES. 143

polar of the point, with respect to a circle whose centre is the origin, and radius k, will be $\dfrac{k^2}{x}$, $\dfrac{k^2}{y}$ respectively. These intercepts completely determine the position of the line, and their *reciprocals* may be taken as its co-ordinates, and denoted by the letters ξ, η.

13. In this system, every equation of the first degree represents a point.

Let
$$a\xi + b\eta = 1$$
be an equation of the first degree.

Draw the straight lines OX, OY at right angles to one another; on OX take the point A, such that $OA = a$, and on OY take the point B, such that $OB = b$. Draw AP, BP perpendicular to OX, OY respectively, meeting in P.

Then, the equation
$$a\xi + b\eta = 1$$
shall represent the point P.

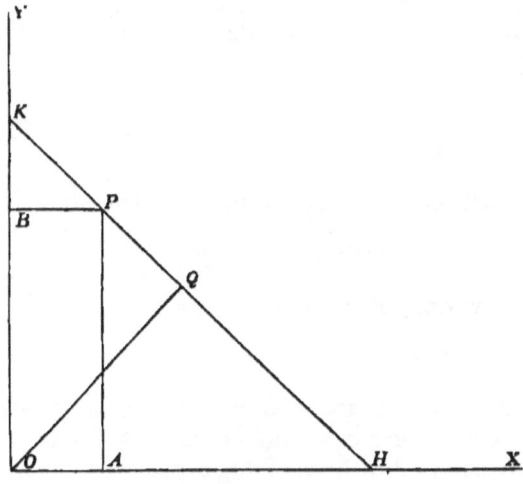

Fig. 2.

Through P draw any straight line meeting OX, OY in H, K, respectively. Then, if ξ, η be the co-ordinates of this line,

$$\frac{1}{OH} = \xi, \quad \frac{1}{OK} = \eta.$$

Hence
$$a\xi = \frac{OA}{OH} = \frac{KP}{HK},$$

$$b\eta = \frac{OB}{OK} = \frac{HP}{HK};$$

$$\therefore a\xi + b\eta = 1;$$

a relation which is satisfied by the co-ordinates of every line passing through the point P. This equation therefore represents the point P.

14. In this system, as in that described in the former part of the present chapter, an equation represents the curve, the co-ordinates of whose tangents satisfy it, and an equation of the nth degree will therefore represent a curve of the nth class.

15. If the perpendicular OQ let fall from O on the straight line HK (fig. 23) be denoted by p, and the angle QOX by ϕ, we shall have

$$\xi = \frac{\cos \phi}{p}, \quad \eta = \frac{\sin \phi}{p},$$

and a point will then be represented by the equation

$$a \cos \phi + b \sin \phi = p;$$

an equation which, if $a^2 + b^2$ be put $= c^2$, and $\frac{b}{a} = \tan \alpha$, becomes $p = c \cos (\phi - \alpha)$.

We thus obtain a method of representing curves by a relation between the perpendicular from a fixed point on the tangent and the inclination of that perpendicular to a fixed straight line. These may be called the *tangential polar co-*

ordinates of the curve. This method will be found discussed in the *Quarterly Journal of Pure and Applied Mathematics*, Vol. I. p. 210.

EXAMPLES.

1. Prove that the distance between the points $a\xi + b\eta = 1$, $a'\xi + b'\eta = 1$, is $\{(a'-a)^2 + (b'-b)^2\}^{\frac{1}{2}}$.

2. Prove that the cosine of the angle between the lines (ξ, η), (ξ', η') is $\dfrac{\xi\xi' + \eta\eta'}{(\xi^2 + \eta^2)^{\frac{1}{2}} (\xi'^2 + \eta'^2)^{\frac{1}{2}}}$.

3. Prove that the distance from the point $(a\xi + b\eta = 1)$ to the line (ξ_1, η_1) is $(a\xi_1 + b\eta_1 - 1)(\xi_1^2 + \eta_1^2)^{-\frac{1}{2}}$.

4. Prove that the equation $\xi^2 + \eta^2 + 2P\xi + 2Q\eta + R = 0$ represents a conic, of which the focus is the origin.

What are the co-ordinates of its directrix? What is its eccentricity, and what its latus-rectum?

5. Prove that the equation $p = a + c\cos\phi$ represents a circle; and determine the radius of the circle.

6. Prove that the evolute of the ellipse $a^2\xi^2 + b^2\eta^2 = 1$ is represented by the equation

$$\frac{a^2}{\xi^2} + \frac{b^2}{\eta^2} = (a^2 - b^2)^2.$$

F.

CHAPTER VIII.

ON THE INTERSECTION OF CONICS, AND ON PROJECTIONS.

1. WE shall here say a few words on the subject of the intersection of two conics, as an acquaintance with this branch of the subject will be useful in future investigations.

Since every conic is represented by an equation of the second degree, it follows that any two conics intersect in four points, which may be (1) all real, (2) two real and two imaginary, or (3) all imaginary.

2. Through these four points of intersection three pairs of straight lines can be drawn. If the four points be called P, Q, R, S, the pairs of straight lines will be PQ and RS, PR and QS, PS and QR. If PQ and RS intersect in L, PR and QS in M, PS and QR in N, the points L, M, N are called (see Art. 15, Chap. II.) the vertices of the quadrangle $PQRS$. Also the three points L, M, N will form, with respect to every conic passing through the points P, Q, R, S, a conjugate triad; and therefore each of them will have the same polar with respect to all such conics.

3. The equations of the pairs of lines PQ, RS, &c. (the sides and diagonals of the quadrangle) may be found as follows. Let the equations of the conics be

$$\phi(\alpha, \beta, \gamma) = u\alpha^2 + v\beta^2 + w\gamma^2 + 2u'\beta\gamma + 2v'\gamma\alpha + 2w'\alpha\beta = 0 \ldots(1),$$
$$\psi(\alpha, \beta, \gamma) = p\alpha^2 + q\beta^2 + r\gamma^2 + 2p'\beta\gamma + 2q'\gamma\alpha + 2r'\alpha\beta = 0 \ldots(2);$$

then every conic passing through their four points of intersection will be represented by an equation of the form

$$\phi(\alpha, \beta, \gamma) + k\psi(\alpha, \beta, \gamma) = 0 \ldots\ldots\ldots\ldots(3).$$

INTERSECTION OF CONICS. 147

If the left-hand member of this equation break up into two factors, the conic degenerates into two straight lines, real or imaginary. The condition that this should happen is

$$\begin{vmatrix} u+pk, & w'+r'k, & v'+q'k \\ w'+r'k, & v+qk, & u'+p'k \\ v'+q'k, & u'+p'k, & w+rk \end{vmatrix} = 0 \quad \ldots \ldots (4),$$

a cubic for the determination of k, of which the roots are either all real, or one real and two imaginary. If the roots be all real, the vertices of the quadrangle, which will be the centres of the several conics included in the form (3), will be all real. If one root only be real, then one vertex only of the quadrangle will be real. We proceed to consider how the reality of the vertices L, M, N depends upon that of the points P, Q, R, S.

4. First, suppose all the four points P, Q, R, S to be real, then it is clear that all the vertices will be real.

5. Next, let two of the points, P, Q, for example, be real, and R, S, imaginary.

Then, the line PR can have no other real point but P. For, if it had, it would itself become a real line, and we should have a real line cutting a real conic in one real and one imaginary point, which is impossible.

Hence the point M, which lies on PR, is imaginary. Similarly the line PS, and the point N, which lies on it, are imaginary. The real vertex must therefore be L, which lies on PQ.

We may observe that the line RS will be real. For the two lines PQ, RS, considered as one locus, will be represented by equation (3) when for k is substituted the real value corresponding to the point L. Hence the form of the expression $\phi(\alpha, \beta, \gamma) + k\psi(\alpha, \beta, \gamma)$ answering to PQ, contains one real linear factor, and the other linear factor, which answers to RS, will therefore also be real.

6. Thirdly, let all the four points of intersection be imaginary. Then the three vertices will all be real.

For, by what has been shewn above, one vertex is necessarily so. Take this as the angular point A of the triangle of reference, and let its polar with respect to the two conics be taken for the side BC.

The point B being chosen arbitrarily, let its polar with respect to one of the conics be taken as AC. Then this conic may be represented by the equation

$$\alpha^2 + v\beta^2 + w\gamma^2 = 0 \quad \ldots\ldots\ldots\ldots\ldots (1).$$

Let the other be represented by

$$\alpha^2 + q\beta^2 + r\gamma^2 + 2p'\beta\gamma = 0 \quad \ldots\ldots\ldots\ldots (2).$$

Since the four points of intersection are imaginary, the roots of the quadratic

$$(q-v)\beta^2 + (r-w)\gamma^2 + 2p'\beta\gamma = 0$$

will be imaginary. Hence

$$(q-v)(r-w) > p'^2 \quad \ldots\ldots\ldots\ldots\ldots (3).$$

Now let $(0, g, h)$ be the co-ordinates of either vertex. Then, since it has the same polar with respect to both conics, the equations

$$vg\beta + wh\gamma = 0,$$
$$(qg + p'h)\beta + (p'g + rh)\gamma = 0,$$

will represent the same straight line, hence

$$\frac{qg + p'h}{vg} = \frac{p'g + rh}{wh}.$$

The two values of $\frac{g}{h}$, given by this equation, will determine the vertices. Now the roots of this equation are real or imaginary, as

$$(qw - rv)^2 + 4vwp'^2 \gtrless 0.$$

VERTICES OF A QUADRANGLE. 149

That is, the vertices will necessarily be real, if v and w have the same signs. Suppose them, however, to have contrary signs, then, by (3),

$$qr - p'^2 > qw + rv - vw\ ;$$

therefore multiplying both sides by $-4vw$, which is a positive quantity,

$$-4vw\,(qr - p'^2) > 4vw\,(vw - qw - rv)\ ;$$

$$\therefore (qw + rv)^2 - 4vw\,(qr - p'^2) > (qw+rv)^2 + 4v^2w^2 - 4vw(qw+rv)$$

$$> (qw + rv - 2vw)^2$$

$$> 0\ ;$$

$$\therefore (qw - rv)^2 + 4vwp'^2 > 0.$$

Hence, when the four points of intersection are imaginary, the vertices are in all cases real.

7. These vertices form, with respect to the two conics, a conjugate triad. Suppose now that they are taken as angular points of the triangle of reference. Let the conics be represented by the equations

$$u\alpha^2 + v\beta^2 + w\gamma^2 = 0,$$

$$p\alpha^2 + q\beta^2 + r\gamma^2 = 0.$$

Then $\pm \dfrac{\alpha}{(qw - rv)^{\frac{1}{2}}} = \pm \dfrac{\beta}{(ru - pw)^{\frac{1}{2}}} = \pm \dfrac{\gamma}{(pv - qu)^{\frac{1}{2}}}$

are the equations of the several pairs of common chords of the two conics. Since two of the expressions

$$qw - rv,\ \ ru - pw,\ \ pv - qu,$$

must necessarily have the same sign, it follows that one pair at least of common chords is always real. The other two pairs will, as may easily be seen, be real or imaginary, according as the four points of intersection are, or are not, all real.

8. Returning to the equation (4) given in Art. 3 of this chapter we see that it may be partially developed into the following form:

$$\begin{vmatrix} u, & w', & v' \\ w', & v, & u' \\ v', & u', & w \end{vmatrix}$$

$$+ \left\{ \begin{vmatrix} p, & w', & v' \\ r', & v, & u' \\ q', & u', & w \end{vmatrix} + \begin{vmatrix} u, & r', & v' \\ w', & q, & u' \\ v', & p', & w \end{vmatrix} + \begin{vmatrix} u, & w', & q' \\ w', & v, & p' \\ v', & u', & v \end{vmatrix} \right\} k$$

$$+ \left\{ \begin{vmatrix} u, & v', & q' \\ w', & q, & p' \\ v', & p', & r \end{vmatrix} + \begin{vmatrix} p, & w', & q' \\ r', & v, & p' \\ q', & u', & r \end{vmatrix} + \begin{vmatrix} p, & r', & v' \\ r', & q, & u' \\ q', & p', & w \end{vmatrix} \right\} k^2$$

$$+ \begin{vmatrix} p, & r' & q' \\ r', & q, & p' \\ q', & p', & r \end{vmatrix} k^3 = 0.$$

This equation is generally written

$$\Delta + \Theta k + \Theta' k^2 + \Delta' k^3 = 0.$$

Here Δ and Δ' are known to be the discriminants of the two given conics. Now the roots of this equation are the values of k which will give the three pairs of straight lines drawn through the points of intersection of the two conics, and since these values must remain unaltered by any transformation of co-ordinates, it follows that the ratios of the four coefficients, Δ, Θ, Θ', Δ', also remain unaltered by any such transformation. On this account they are called the *Invariants* of the system. They possess numerous interesting properties, but a detailed examination of them would lead us too far from the object of this work. They will be found fully discussed in Dr Salmon's treatise. As an example of their use, however, we may demonstrate the following proposition:

If $S=0$, $S'=0$ be the equations of two conics, $\Sigma=0$ that of the reciprocal of the first with respect to the second, $\Sigma'=0$ that of the second with respect to the first, then

$$\Sigma - \Theta S' + \Theta' S - \Sigma' = 0$$

identically.

The direct investigation of this, taking the equations in their most general form, will be somewhat laborious. Suppose, however, that the conics, referred to their common conjugate triad, are represented by the equations

$$S = u\alpha^2 + v\beta^2 + w\gamma^2 = 0 \ldots\ldots(S),$$
$$S' = p\alpha^2 + q\beta^2 + r\gamma^2 = 0 \ldots\ldots(S').$$

Then $\Delta = uvw$, $\Theta = pvw + qwu + ruv$,

$\Delta' = pqr$, $\Theta' = uqr + vrp + wpq$.

Also $\Sigma = \begin{vmatrix} 0, & p\alpha, & q\beta, & r\gamma \\ p\alpha, & u, & 0, & 0 \\ q\beta, & 0, & v, & 0 \\ r\gamma, & 0, & 0, & w \end{vmatrix} = p^2 vw\alpha^2 + q^2 wu\beta^2 + r^2 uv\gamma^2.$

And $\Sigma' = u^2 qr\alpha^2 + v^2 rp\beta^2 + w^2 pq\gamma^2,$

$\Theta' S - \Theta S' = (uqr + vrp + wpq)(u\alpha^2 + v\beta^2 + w\gamma^2)$
$\qquad - (pvw + qwu + ruv)(p\alpha^2 + q\beta^2 + r\gamma^2)$
$\qquad = (u^2 qr - p^2 vw)\alpha^2 + (v^2 rp - q^2 wu)\beta^2 + (w^2 pq - r^2 uv)\gamma^2$
$\qquad = \Sigma' - \Sigma.$

Hence, as stated above, $\Sigma - \Theta S' + \Theta' S - \Sigma' = 0$ *identically*, in the case where the conics are referred to their common conjugate triad. And, from what has been said above, it appears that on account of the invariance of Θ and Θ', the same result holds in whatever manner the conics may be expressed. We thus see that if each of two conics be reciprocated with respect to the other, the four points of intersection of any two of the conics thus obtained, and the four points of intersection of the other two, lie on a conic.

On Projections.

9. DEF. The surface generated by a straight line of indefinite length, which always passes through a given fixed point, and always meets a given curve, the curve and point not lying in the same plane, is called a *cone*.

The fixed point is called the *vertex*, and will be denoted in this chapter by the letter V.

If a cone be cut by any two planes, either of the curves of section is said to be a *projection* of the other.

Also the two points in which any generating line is cut by two planes are said to be the projections, the one of the other. The straight line in which the two planes intersect is called the *Unprojected*.

It may easily be seen that the projection of any curve on a given plane coincides with the shadow of the curve which would be cast upon the plane by a luminous point coinciding with the vertex of the cone.

The projection of a point of intersection of any two curves will be a point of intersection of their projections.

The projection of any straight line will be a straight line; and that of any curve of the nth degree will be a curve of the nth degree. For since any straight line and curve of the nth degree intersect in n points, their projections will also intersect in n points.

10. If AB be any given straight line, and a cone be cut by any plane parallel to VAB, the projection of the line AB will be infinitely distant. Hence it is always possible so to project a figure, that the projection of any given straight line shall be removed to an infinite distance. This is called projecting the straight line to infinity.

11. Any quadrilateral may be projected into a parallelogram.

For, if $ABCD$ be any quadrilateral, and the sides AB, CD be produced to meet in E, AD, BC in F, and the line EF projected to infinity, then, since the projections of AB,

PROJECTION OF TED INTO A CIRCLE. 155

CD intersect at an infinite distance, which *V* is the common one another, as also those of *AD*, *BC*, conics. The sections the quadrilateral *ABCD* is projected into to the plane *VAB* are parallel to

12. The angle *EVF* will be the angle between lar and similar jections of the sides *AB*, *BC*. For if the plane form. tion cut the lines *VA*, *VB*, *VC*, *VD* in *A'*, *B'*, *C*, spectively, then the points *A'*, *B'*, *C'*, *D'* are respectively different projections of *A*, *B*, *C*, *D*. Now the plane *ABA'B'* contains, the points *V*, *E*, and, since the plane of projection, in which the points *A'*, *B'* lie, is parallel to *VEF*, and therefore to *VE*, it follows that *A'B'* is parallel to *VE*. Similarly *B'C'* is parallel to *VF*, and therefore the angle *A'B'C'* is equal to the angle *EVF*.

13. Since the angle *EVF* may be made of any magnitude, by taking the point *V* anywhere on any segment of a circle of which *EF* is the base and which contains an angle of the required magnitude, it follows that any quadrilateral may be projected, in an infinite number of ways, into a parallelogram of which the angles are of any assigned magnitude.

14. We may now proceed to detail the application of the theory of projections to curves of the second degree.

It will easily be seen that the projection of any tangent to any curve will be a tangent to the projection of the curve.

Again, if any point and straight line be the pole and polar of one another with respect to a given conic, their projections will be the pole and polar of one another with respect to the projection of the conic.

For, let *O* be any given point, *XY* its polar with respect

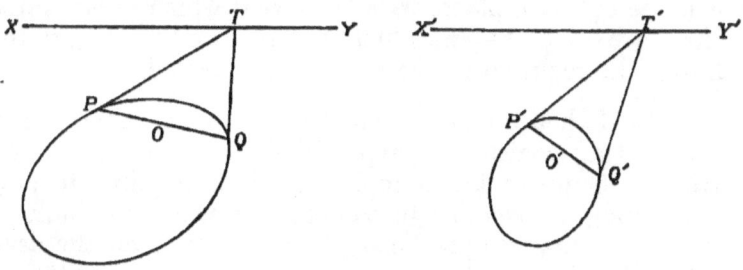

Fig. 24.

to any given conic. On XY take any point T, external to the conic, and from T draw two tangents TP, TQ, then PQ will pass through O. Now project the whole system, and let O', P', Q', T', X', Y' be the respective projections of O, P, Q, T, X, Y. Then $T'P'$, $T'Q'$ will be tangents to the projected conic, and $P'Q'$ will pass through O'. Hence since T is any point on $X'Y'$, $X'Y'$ will be the polar of O'.

15. From the proposition just proved, it will follow that any two conics may be projected into concentric curves. For it is always possible (Arts. 5 and 7) to find one real point at least, the polar of which with respect to two given conics is the same straight line. Let then this straight line be projected to infinity, and its common pole, with respect to the two conics, will become the centre of the curves of projection.

16. It may also be proved that any two conics may be projected into similar and similarly situated curves. For it is always possible (Arts. 5 and 7) to find two straight lines which meet two given conics in the same two points, real or imaginary. Project either of these straight lines to infinity, and the conics will then be projected into curves, two of the points of intersection of which are infinitely distant, that is, into similar and similarly situated conics. These will be ellipses or hyperbolas, according as the points, in which the line projected to infinity meets the conics, are imaginary or real. If the two conics have double contact with one another, their projections will also be concentric.

17. The projections, spoken of in the last two articles, may be effected in an infinite number of ways. For any point whatever may be taken as the vertex of the cone, and if the cone be cut by a plane, parallel to that which passes through the vertex and the line which it is required to project to infinity, the required projection will be effected.

18. It hence follows that it is possible to project any two intersecting conics into hyperbolas of any assigned eccentricity. Suppose, for example, that it is required to project two conics, intersecting in points A, B, into two similar and similarly situated hyperbolas, the angle between the asymptotes of each being a. Take any point V, such that the angle

$AVB = \alpha$, and describe two cones, of which V is the common vertex, passing through the two given conics. The sections of these cones made by any plane parallel to the plane VAB will be hyperbolas, of which the asymptotes are parallel to VA, VB respectively, and will therefore be similar and similarly situated to one another, and of the required form.

19. We now come to the most important and most difficult point of the theory of projections, the process by which, from the properties of the circle, those of conic sections in general may be deduced. We have just seen that any two conics may be projected into hyperbolas of any assigned eccentricity. Now this process, the possibility of which we have shewn by a geometrical method, of course admits of algebraical proof. And the algebraical investigation, on account of the continuity of the symbols employed, would not take any account of the restrictions introduced into the geometrical investigation, either as to the conics intersecting in real points, or as to the eccentricity of the conics into which they are projected being greater than unity. It is therefore possible, by an algebraical process, to transform the equations of *any two conics whatever* into those of conics of *any eccentricity*, and therefore into those of circles. The points and tangents common to the two given conics will be transformed into points and tangents common to their projections, and the relations of poles and polars will remain unaltered.

Since all circles pass through the same two points on the line at infinity, it follows that all circles are transformed by projection into a system of conics passing through the same two points, or having a common chord. Again, since every parabola touches the line at infinity, it follows that all parabolas will project into a system of conics touching the same straight line. A system of parabolas and circles will project into a system in which all the circles will become conics passing through the same two points, and all the parabolas will become conics, having the straight line joining those two points for a common tangent.

20. We have seen, in the investigation of the co-ordinates of the real and imaginary foci, given in Chap. VI., that the pair

of imaginary tangents, drawn to a conic from any one of its four foci, satisfy the analytical conditions of being asymptotes to a circle. Hence these tangents must themselves meet the line at infinity in the two circular points. Conversely, if from the two circular points at infinity two pairs of tangents be drawn to any conic, these will form an imaginary quadrilateral, circumscribing the conic, the four angular points of which are the four foci of the curve.

Hence all conics having the same focus project into conics having a pair of common tangents; and all confocal conics into conics inscribed in the same quadrilateral.

The directrix is the polar of the focus, hence, if two conics have the same focus and directrix, they project into two conics having a common chord of contact for their common tangents, that is, having double contact with one another.

21. *The anharmonic ratio of any pencil or range is unaltered by projection.*

Let the transversal $PQRS$ cut the four straight lines OP, OQ, OR, OS. Take any point V, not lying in the plane through these straight lines, join VO, VP, VQ, VR, VS, and let these lines be cut by any other plane in O', P', Q', R', S'. Then

$$\{O' \cdot P'Q'R'S'\} = [P'Q'R'S']$$

$$= \frac{P'Q' \cdot R'S'}{P'R' \cdot Q'S'}$$

$$= \frac{\sin P'VQ' \cdot \sin R'VS'}{\sin P'VR' \cdot \sin Q'VS'}$$

$$= \frac{\sin PVQ \cdot \sin RVS}{\sin PVR \cdot \sin QVS}$$

$$= \frac{PQ \cdot RS}{PR \cdot QS}$$

$$= [PQRS]$$

$$= \{O \cdot PQRS\}.$$

PROJECTION OF ANGLES.

Hence the anharmonic ratio of the given pencil and range is the same as that of their projection.

22. The following proposition is useful in the projection of theorems relating to the magnitude of angles.

Any two lines which make an angle A *with each other, form with the lines joining the circular points at infinity to their point of intersection, a pencil of which the anharmonic ratio is* $\epsilon^{(\pi-2A)\sqrt{-1}}$.

It will be understood that the two given lines are taken as the first and third legs of the pencil.

Take the two lines as two sides of the triangle of reference, and let them be denoted by $\beta = 0$, $\gamma = 0$. The lines joining their point of intersection to the circular points at infinity are given by eliminating α between the equation of the line at infinity and that of the circumscribing circle, that is, between

$$a\alpha + b\beta + c\gamma = 0, \quad \frac{a}{\alpha} + \frac{b}{\beta} + \frac{c}{\gamma} = 0.$$

This gives $\quad \beta^2 + 2\beta\gamma \cos A + \gamma^2 = 0.$

Now the two lines represented by the equation

$$(\beta - k\gamma)(\beta + k'\gamma) = 0$$

form with $\beta = 0$ and $\gamma = 0$ a pencil of which the anharmonic ratio is $\dfrac{k'}{k}$ (Art. 23, Chap. I.). In the present case,

$$k = -\epsilon^{A\sqrt{-1}}, \quad k' = \epsilon^{-A\sqrt{-1}}.$$

Hence the anharmonic ratio is

$$-\frac{\epsilon^{A\sqrt{-1}}}{\epsilon^{-A\sqrt{-1}}} = -\epsilon^{2A\sqrt{-1}} = \epsilon^{(\pi-2A)\sqrt{-1}}.$$

COR. In the case in which the lines are at right angles to one another, $A = \dfrac{\pi}{2}$, and the anharmonic ratio becomes unity, that is, the four lines form an harmonic pencil.

23. The known property of a circle, that "the angles in the same segment are equal to one another," gives rise to an important anharmonic property of conic sections. The property of the circle may be expressed thus, that "if A, B be any two fixed points on the circumference of a circle, O any moving point on it, the angle AOB is constant." Project the circle into any conic, and let A', B', O' be the projections of A, B, O; H, K those of the circular points at infinity. Then, from the result of the last article, it follows that

$$\{O' \cdot A'B'HK\} \text{ is constant.}$$

Or, *the anharmonic ratio of the pencil, formed by joining any point of a conic to four fixed points on the curve, is constant.*

Reciprocating this theorem, in accordance with Art. 13, Chap. VII., we see that *if any tangent to a conic be cut by four fixed tangents, the anharmonic ratio of the range, formed by the points of section, is constant.*

24. If P, Q, R be three points in a straight line, and p, q, r be their projections, and s the projection of the point at infinity on the line PQR, then

$$[pqrs] = \frac{PQ}{QR}.$$

For $$[pqrs] = \frac{pq \cdot rs}{ps \cdot qr} = \frac{PQ}{PS} \cdot \frac{RS}{QR},$$

where S denotes the point at infinity on the line PQR.

Also $RS : PS$ in a ratio of equality, hence

$$[pqrs] = \frac{PQ}{QR}.$$

25. If $P, P', Q, Q', R, R'\ldots$ be a system of points in involution, and $p, p', q, q', r, r'\ldots$ their projections, then since by Art. 27, Chap. I. $[PQRS] = [P'Q'R'S']$, and by Art. 21 of this Chapter $[PQRS] = [pqrs]$, $[P'Q'R'S'] = [p'q'r's']$,

it follows that $[pqrs] = [p'q'r's']$, or $p, p', q, q', r, r'...$ are a system of points in involution. Hence, *any system of points in involution projects into a system in involution.*

If P coincides with P', p will coincide with p', or the foci of one system project into the foci of the other. We may observe that the centre of one system will *not*, in general, project into the centre of the other.

26. Let a system of circles be described through two given points A, A', and let any circle of the system cut a given straight line in P, P'. Produce AA' to meet the given straight line in O. Then

$$OP . OP' = OA . OA',$$

or $OP . OP'$ is constant for all circles passing through A, A'. Hence, the system of points in which a system of circles, passing through two given points, cut a given straight line, are in involution. Project the system of circles into a system of conics, passing through four given points, and we learn that "a system of conics, passing through four given points, cut any straight line in a system of points in involution."

Of this system of conics, one can be drawn so that one of its points of intersection with the given straight line shall be at an infinite distance,—in other words, so that one of its asymptotes shall be parallel to the given straight line. The other point, in which this conic cuts the given straight line, will be the centre of the system.

Again (see Art. 3, Chap. IX., *infra*), *two* conics can be described, passing through the four given points, and touching the given straight line. The two points of contact of these conics will be the foci of the system of points in involution.

By reciprocating these propositions, we obtain analogous properties of the system of conics, inscribed in a given quadrilateral, whence, by projection, may be obtained those of a system of confocal conics.

27. When the vertex of the cone, used for purposes of projection, is infinitely distant, so that the cone itself be-

comes a cylinder, the projection is said to be *orthogonal*. In this mode of projection, the line at infinity remains at an infinite distance, and any two parallel lines will therefore project into parallel lines. Also any area will bear to its projection a constant ratio; and the mutual distances of any three points in the same straight line will bear to one another the same ratios as the mutual distances of their projections. Two perpendicular diameters of a circle will, since each is parallel to the tangent at the extremity of the other, project into two conjugate diameters of an ellipse. By this method, many properties of conic sections, more especially those relating to conjugate diameters, may be readily deduced from those of the circle.

EXAMPLES.

1. If XYZ be a triangle which moves in such a manner that its side YZ always passes through a fixed point P, ZX through Q, XY through R, and if the locus of Y be a fixed conic passing through R and P, that of Z a fixed conic passing through P and Q, prove that the locus of X will be a fixed conic passing through Q, R, and through the other three points of intersection of the two given conics.

2. If two tangents be drawn to a conic so that the points in which they cut a given straight line form, with two fixed points on the straight line, a harmonic range, prove that the locus of their point of intersection will be a conic passing through the two given points.

3. A system of conics is described touching four given straight lines; prove that the locus of the pole of any fifth given straight line with respect to any conic of the system is a straight line.

If the fifth straight line be projected to infinity so that the points where it intersects two of the other given straight lines be projected into the circular points, what does this theorem become?

EXAMPLES.

4. A system of conics is described about a given quadrangle; prove that the locus of the pole of any given straight line, with respect to any conic of the system, is a conic passing through the vertices of the quadrangle.

5. A system of conics is described touching the sides of a given triangle, and from a given point a pair of tangents is drawn to each conic of the system. Prove that, if the locus of one of the points of contact be a straight line, that of the other will be a conic circumscribed about the given triangle.

6. The tangent at any point P of a conic, of which S and H are the foci, is cut by two conjugate diameters in T, t; prove that the triangles SPT, HPt are similar to one another.

7. Two parabolas have a common vertex and their axes at right angles to each other; prove that the polar reciprocal of either with respect to the other is a rectangular hyperbola, of which the axes of the parabolas are the asymptotes.

CHAPTER IX.

MISCELLANEOUS PROPOSITIONS.

ON THE DETERMINATION OF A CONIC FROM FIVE GIVEN GEOMETRICAL CONDITIONS.

1. IF any five independent conditions be given, to which a conic is to be subject, each of these, expressed in algebraical language, will give an equation for the determination of the five arbitrary constants which the equation of the conic involves. Hence, five conditions suffice for the determination of the conic. It may, however, happen that some of the equations for the determination of the constants rise to a degree higher than the first, in such a case, the constants will have more than one value, and more than one conic may therefore be described, satisfying the required conditions, although the number will still be finite.

The geometrical conditions of most frequent occurrence are those of passing through given points and touching given straight lines, with such others as may be reduced to these. We proceed to consider how many conics may be described in each individual case.

2. *Let five points be given.*

In this case we have merely to substitute in the equation of the conic the co-ordinates of the several points for a, β, γ; we shall thus obtain five simple equations for the determination of the constants, and one conic only will satisfy the given conditions.

3. *Let four points and one tangent be given.*

Take three of the points as angular points of the triangle of reference. Let f, g, h be the co-ordinates of the fourth

DETERMINATION OF A CONIC FROM FIVE CONDITIONS. 163

given point, $l\alpha + m\beta + n\gamma = 0$, the equation of the given tangent. Let the equation of the conic be

$$\frac{\lambda}{\alpha} + \frac{\mu}{\beta} + \frac{\nu}{\gamma} = 0.$$

Then for the determination of the ratios $\lambda : \mu : \nu$, we have the equations

$$\frac{\lambda}{f} + \frac{\mu}{g} + \frac{\nu}{h} = 0,$$

$$\lambda^2 l^2 + \mu^2 m^2 + \nu^2 n^2 - 2\mu\nu mn - 2\nu\lambda nl - 2\lambda\mu lm = 0.$$

These equations will give two values for the ratios, and prove therefore that two conics can be described satisfying the required conditions.

4. *Let three points and two tangents be given.*

Take the three points as angular points of the triangle of reference. Let the two given tangents be represented by the equations

$$l\alpha + m\beta + n\gamma = 0,$$
$$l'\alpha + m'\beta + n'\gamma = 0.$$

If then the conic be represented by the equation

$$\frac{\lambda}{\alpha} + \frac{\mu}{\beta} + \frac{\nu}{\gamma} = 0,$$

we have, for the determination of $\lambda : \mu : \nu$, the equations

$$\lambda^2 l^2 + \mu^2 m^2 + \nu^2 n^2 - 2\mu\nu mn - 2\nu\lambda nl - 2\lambda\mu lm = 0,$$
$$\lambda^2 l'^2 + \mu^2 m'^2 + \nu^2 n'^2 - 2\mu\nu m'n' - 2\nu\lambda n'l' - 2\lambda\mu l'm' = 0,$$

which, being both quadratics, give four values for each of the ratios, shewing that four conics may be described satisfying the given conditions.

5. *Let two points and three tangents be given.*

Take the three tangents as lines of reference, and let f, g, h; f', g', h', be the co-ordinates of the two given points.

Then, if the equation of the conic be

$$\lambda^2\alpha^2 + \mu^2\beta^2 + \nu^2\gamma^2 - 2\mu\nu\beta\gamma - 2\nu\lambda\gamma\alpha - 2\lambda\mu\alpha\beta = 0,$$

we shall get, writing f, g, h; f', g', h', successively for α, β, γ, two quadratics for the determination of the ratios $\lambda : \mu : \nu$, giving therefore four conics.

6. *Let one point and four tangents be given.*

Taking three of the tangents as lines of reference, the condition of touching the fourth given line gives a simple equation for the determination of the coefficients, and that of passing through the given point a quadratic. Hence, two conics may be described, satisfying the given conditions.

7. *Let five tangents be given.*

Taking three of the tangents as lines of reference, the condition of touching each of the others gives a simple equation for the determination of the constants, shewing that one conic only can be described satisfying these conditions.

The results of Arts. 5, 6, 7, may of course be deduced by the method of reciprocal polars, from those of Arts. 4, 3, 2.

8. Several other forms under which the data may be given, are reducible to a certain number of lines and points. Thus to have given a tangent and its point of contact is equivalent to having two points given, the points being indefinitely close together. Or, it may be regarded as equivalent to having two tangents given, these tangents being indefinitely nearly coincident. To have given that a conic is a parabola is equivalent to having a tangent given, since every parabola touches the line at infinity. To have given that it is a circle is equivalent to having two points given, since all circles intersect the line at infinity in the same two points. And this explains the reason why four circles can be described touching the sides of a given triangle, but only one circumscribed about it. So, to have given that a conic is similar and similarly situated to a given one is equivalent to having two points given. To have given an asymptote is equivalent to having two points given, for an asymptote may be regarded as a tangent, the point of contact of which is given (at an infinite distance). To have given the direction of an asymptote is equivalent to having one point given, for this virtually determines the point in which the conic meets the line at infinity.

9. If it be given that three given points form a conjugate triad, this is equivalent to three conditions, as the equation of the conic, when these are taken as angular points of the triangle of reference, is of the form

$$u\alpha^2 + v\beta^2 + w\gamma^2 = 0.$$

Two more conditions will therefore completely determine the conic. If these conditions be that the conic shall pass through two given points, or touch two given straight lines, or pass through one given point and touch one given straight line, one conic only can be drawn to satisfy these conditions.

We may observe that, if the above conic pass through the point (f, g, h), it also passes through the three points $(-f, g, h)$, $(f, -g, h)$, $(f, g, -h)$, and that, if it touch the line (l, m, n), it also touches the lines $(-l, m, n)$, $(l, -m, n)$, $(l, m, -n)$.

ON THE LOCUS OF THE CENTRE OF A SYSTEM OF CONICS WHICH SATISFY FOUR CONDITIONS, EXPRESSED BY PASSING THROUGH POINTS AND TOUCHING STRAIGHT LINES.

10. The locus of the centre of a conic, which passes through m points, and touches n straight lines, $m + n$ being equal to four, will be a conic, in every case except two. We will consider the several cases in order.

11. Let the system pass through four points.

This is best treated by Cartesian co-ordinates.

Of the conics which can be described passing through four points, two are parabolas. Take, as co-ordinate axes, that diameter of each of these parabolas, the tangent at the extremity of which is parallel to the axis of the other. Then the two parabolas will be represented by the equations

$$x^2 + 2fy + h = 0 \quad \text{...................(1)},$$
$$y^2 + 2g'x + h' = 0 \quad \text{...................(2)}.$$

The system of conics is represented by the equation

$$x^2 + \lambda y^2 + 2\lambda g'x + 2fy + h + \lambda h' = 0 \quad \text{.........(3)},$$

λ being an arbitrary multiplier.

The centre is given by the equations

$$x + \lambda g' = 0, \quad \lambda y + f = 0.$$

Eliminating λ, we get for the locus of the centres

$$xy = fg \quad \text{........................(4)},$$

a conic, whose asymptotes are parallel to the axes of the parabolas (1) and (2).

If the four points form a convex quadrangle, the parabolas will be real, and the locus (4) an hyperbola. If the quadrangle be concave, the parabolas will be imaginary, and the locus of the centres an ellipse.

The curve (4) bisects the distance between each pair of the four points, and passes through the vertices of the quadrangle. This may be seen from geometrical considerations, for the three pairs of straight lines which belong to this system of conics, the vertices are respectively the centres.

From the form of the equation (3) we see that every conic of the system has a pair of conjugate diameters parallel to the axes of the parabola (1), (2); in other words to the asymptotes of (4).

The conic of minimum eccentricity is obtained by making $\lambda = 1$. In this case, these are the equal conjugate diameters.

If the axes of the parabolas be at right angles to one another, the four points lie on the circumference of a circle. The axes of every conic in (3) are then parallel to the coordinate axes, and (4) is a rectangular hyperbola.

If each of the four points be the orthocentre of the other

three, the system of conics is a system of rectangular hyperbolas, and (4) is the nine-point circle of the given points.

12. *Let three points and a tangent be given.*

In this case we may see, *à priori*, that the locus will be a curve of the fourth degree, for we can describe four parabolas satisfying the given conditions, and the locus will have four asymptotes, parallel to the axes of these parabolas.

Take the triangle formed by the three points for the triangle of reference, and use triangular co-ordinates. Let the tangent be represented by $lx + my + nz = 0$.

Then, if the system of conics be represented by
$$\lambda yz + \mu zx + \nu xy = 0,$$
the condition of tangency gives
$$l^2\lambda^2 + m^2\mu^2 + n^2\nu^2 - 2mn\mu\nu - 2nl\nu\lambda - 2lm\lambda\mu = 0.$$

The centre is given by the equation
$$\mu z + \nu y = \nu x + \lambda z = \lambda y + \mu x.$$

If each member of this be put for the moment $= \rho$, we have
$$\lambda = \frac{y+z-x}{2yz}\rho, \quad \mu = \frac{z+x-y}{2zx}\rho, \quad \nu = \frac{x+y-z}{2xy}\rho;$$
therefore the equation of the locus becomes
$$l^2x^2(y+z-x)^2 + m^2y^2(z+x-y)^2 + n^2z^2(x+y-z)^2$$
$$- 2mnyz(z+x-y)(x+y-z) - 2nl\,lx(x+y-z)(y+z-x)$$
$$- 2lmxy(y+z-x)(z+x-y) = 0,$$
a curve of the fourth degree.

Writing $1-2x$, $1-2y$, $1-2z$, for
$$y+z-x, \quad z+x-y, \quad x+y-z,$$
respectively, the terms of the fourth order become
$$l^2x^2 + m^2y^2 + n^2z^2 - 2mny^2z^2 - 2nlz^2x^2 - 2lmx^2y^2.$$

Hence the asymptotes, and therefore the axes of the four parabolas, are parallel to the four lines

$$\pm l^{\frac{1}{2}}x \pm m^{\frac{1}{2}}y \pm n^{\frac{1}{2}}z = 0.$$

13. *Let two points and two tangents be given.*

In this case, again, four parabolas can be described satisfying the given conditions, and we might therefore expect that the locus would be a curve of the fourth degree. It will be found, however, that it breaks up into two factors of the second degree.

Taking the line joining the two points as $\alpha = 0$, and the other two as $\beta = 0$, $\gamma = 0$, the equation of the system may be written

$$2\beta\gamma + (\lambda\alpha + m\beta + n\gamma)^2 = 0.$$

Here λ is a variable parameter. For the determination of m and n, we may proceed as follows. Let the values of $\dfrac{\beta}{\gamma}$, corresponding to $\alpha = 0$, be called l, l'. We have then

$$(\beta - l\gamma)(\beta - l'\gamma) = \beta^2 + 2\,\frac{mn+1}{m^2}\,\beta\gamma + \frac{n^2}{m^2}\,\gamma^2 \text{ identically,}$$

$$\therefore l + l' = -2\,\frac{mn+1}{m^2},$$

$$ll' = \frac{n^2}{m^2}.$$

Hence $\qquad \dfrac{n}{m} = \pm (ll')^{\frac{1}{2}}.$

For the centre, we have the equations

$$\lambda(\lambda\alpha + m\beta + n\gamma) = m(\lambda\alpha + m\beta + n\gamma) + \gamma$$
$$= n(\lambda\alpha + m\beta + n\gamma) + \beta;$$

$$\therefore \lambda\alpha + m\beta + n\gamma = \frac{\beta - \gamma}{m - n};$$

GIVEN, ONE POINT AND THREE TANGENTS. 169

$$\therefore \lambda = \frac{(m+n)(\lambda\alpha + m\beta + n\gamma) + \beta + \gamma}{2(\lambda\alpha + m\beta + n\gamma)} = \frac{m+n}{2} + \frac{m-n}{2}\frac{\beta+\gamma}{\beta-\gamma}$$

$$= \frac{m\beta - n\gamma}{\beta - \gamma}.$$

Hence, the locus becomes

$$\alpha\lambda\frac{m\beta - n\gamma}{\beta - \gamma} + m\beta + n\gamma = \frac{\beta-\gamma}{m-n},$$

or $(m-n)\alpha(m\beta - n\gamma) + (m-n)(\beta-\gamma)(m\beta + n\gamma) = (\beta-\gamma)^2$;

$\therefore (m-n)\{m\beta^2 - n\gamma^2 + (n-m)\beta\gamma - n\gamma\alpha + m\alpha\beta\} = (\beta-\gamma)^2.$

This may be written

$$(m^2 - mn - 1)\beta^2 + (n^2 - mn - 1)\gamma^2 - (m^2 - 2mn + n^2 - 2)\beta\gamma$$
$$+ (n^2 - mn)\gamma\alpha + (m^2 - mn)\alpha\beta = 0,$$

or, dividing by m^2, and substituting the values of $\frac{n}{m}$ already obtained,

$$\left(1 - \frac{l+l'}{2}\right)\beta^2 + \left(ll' - \frac{l+l'}{2}\right)\gamma^2 - (1 + l + l' + ll')\beta\gamma$$
$$+ ll'\gamma\alpha + \alpha\beta \pm (ll')^{\frac{1}{2}}(\gamma\alpha + \alpha\beta) = 0,$$

giving, as stated above, two conic sections for the required locus.

14. *Let one point, and three tangents be given.*

Here the required locus will be a conic, since only two parabolas can be described satisfying the given conditions.

Take the three straight lines as lines of reference; and let f, g, h, be the triangular co-ordinates of the given point. We have then, as the equation of the system,

$$l^2x^2 + m^2y^2 + n^2z^2 - 2mn\,yz - 2nl\,zx - 2lm\,xy = 0,$$

subject to the condition

$$l^2f^2 + m^2g^2 + n^2h^2 - 2mn\,gh - 2nl\,hf - 2lm\,fg = 0\ldots\ldots(1).$$

For the centre, we have

$$l(lx-my-nz) = m(-lx+my-nz) = n(-lx-my+nz);$$

$$\therefore \frac{lx-ny-nz}{mn} = \frac{-lx+my-nz}{nl} = \frac{-lx-my+nz}{lm};$$

or
$$\frac{lx}{l(m+n)} = \frac{my}{m(n+l)} = \frac{nz}{n(l+m)};$$

$$\therefore \frac{x}{m+n} = \frac{y}{n+l} = \frac{z}{l+m};$$

$$\therefore \frac{l}{y+z-x} = \frac{m}{z+x-y} = \frac{n}{x+y-z},$$

giving, for the locus of the centre,

$$f^2(y+z-x)^2 + g^2(z+x-y)^2 + h^2(x+y-z)^2$$
$$- 2gh(z+x-y)(x+y-z) - 2hf(x+y-z)(y+z-x)$$
$$- 2fg(y+z-x)(z+x-y) = 0,$$

a conic touching the three straight lines which join the middle points of the sides of the triangle formed by the three given tangents.

Its asymptotes are parallel to those of the curve

$$f^2x^2 + g^2y^2 + h^2z^2 - 2ghyz - 2hfzx - 2fgxy = 0.$$

It will therefore be a rectangular hyperbola, if

$$a^2f^2 + b^2g^2 + c^2h^2 + (b^2+c^2-a^2)gh + (c^2+a^2-b^2)hf$$
$$+ (a^2+b^2-c^2)fg = 0 \text{ (Art. 3, Chap. v.),}$$

that is, if

$$(b^2+c^2-a^2)(g+h)^2 + (c^2+a^2-b^2)(h+f)^2$$
$$+ (a^2+b^2-c^2)(f+g)^2 = 0;$$

or, if the point (f, g, h) lie anywhere on the circumference of the circle, which is self-conjugate with respect to the triangle formed by drawing through the intersection of each pair of the three given straight lines, a straight line parallel to the third.

It will be a circle, if
$$\left(\frac{g+h}{a}\right)^2 = \left(\frac{h+f}{b}\right)^2 = \left(\frac{f+g}{c}\right)^2;$$
or, if
$$\frac{f}{b+c-a} = \frac{g}{c+a-b} = \frac{h}{a+b-c} = \frac{1}{a+b+c},$$
$$\frac{-f}{a+b+c} = \frac{g}{a+b-c} = \frac{h}{c+a-b} = \frac{1}{a-b-c},$$
$$\frac{f}{a+b-c} = \frac{-g}{a+b+c} = \frac{h}{b+c-a} = \frac{1}{b-c-a},$$
$$\frac{f}{c+a-b} = \frac{g}{b+c-a} = \frac{-h}{a+b+c} = \frac{1}{c-a-b}.$$

It will be a parabola, if $fgh(f+g+h) = 0$, that is, if the point be at an infinite distance or on any one of the three given straight lines.

15. *Let four tangents be given.*

In this case, but one parabola can be described, and we may anticipate that the locus will be a straight line. This may be proved algebraically as follows. Take the diagonals of the quadrilateral formed by the four given tangents as lines of reference, and let the equations of the four tangents be
$$\pm lx \pm my \pm nz = 0.$$
If the equation of the conic be
$$ux^2 + vy^2 + wz^2 = 0 \quad\ldots\ldots\ldots\ldots\ldots\ldots(1),$$
we must have
$$vwl^2 + wum^2 + uvn^2 = 0 \quad\ldots\ldots\ldots\ldots(2).$$
And the centre of (1) is given by the equations
$$ux = vy = wz;$$
therefore its locus is represented by
$$l^2 x + m^2 y + n^2 z = 0.$$

SUPPLEMENTARY PROBLEMS.

16. The following theorem is useful in many geometrical investigations.

The product of any two determinants is a determinant.

First, take the case of two rows and columns.

Let
$$\left. \begin{array}{l} a_1 x_1 + a_2 x_2 = \xi_1 \\ b_1 x_1 + b_2 x_2 = \xi_2 \end{array} \right\} \dots\dots\dots\dots\dots(1).$$

And let
$$\left. \begin{array}{l} \alpha_1 \xi_1 + \alpha_2 \xi_2 = 0 \\ \beta_1 \xi_1 + \beta_2 \xi_2 = 0 \end{array} \right\} \dots\dots\dots\dots\dots(2).$$

These equations lead to the following:

$$\alpha_1 (a_1 x_1 + a_2 x_2) + \alpha_2 (b_1 x_1 + b_2 x_2) = 0,$$
$$\beta_1 (a_1 x_1 + a_2 x_2) + \beta_2 (b_1 x_1 + b_2 x_2) = 0;$$

or
$$\left. \begin{array}{l} (\alpha_1 a_1 + \alpha_2 b_1) x_1 + (\alpha_1 a_2 + \alpha_2 b_2) x_2 = 0 \\ (\beta_1 a_1 + \beta_2 b_1) x_1 + (\beta_1 a_2 + \beta_2 b_2) x_2 = 0 \end{array} \right\} \dots\dots\dots(3).$$

Now, if (2) be satisfied, (3) will be.

And (2) are satisfied, if either

$$\begin{vmatrix} \alpha_1, & \alpha_2 \\ \beta_1, & \beta_2 \end{vmatrix} = 0 \dots\dots\dots\dots\dots(4),$$

or if ξ_1 and ξ_2 are each $= 0$.

In the latter case, we have by (1), either

$$\begin{vmatrix} a_1, & a_2 \\ b_1, & b_2 \end{vmatrix} = 0 \dots\dots\dots\dots\dots(5),$$

or x_1 and $x_2 = 0$. But if x_1 and x_2 be not $= 0$, then (3) gives

$$\begin{vmatrix} \alpha_1 a_1 + \alpha_2 b_1, & \alpha_1 a_2 + \alpha_2 b_2 \\ \beta_1 a_1 + \beta_2 b_1, & \beta_1 a_2 + \beta_2 b_2 \end{vmatrix} = 0 \dots\dots\dots\dots(6).$$

Hence (6) is satisfied whenever either (4) or (5) are, and therefore its left-hand member must involve, as factors, the left-hand members of (4) and (5). The only other factor is numerical, and this will be seen, by comparing the coefficients of any term, as for instance $\alpha_1 \beta_2 a_1 b_2$, to be unity.

SUPPLEMENTARY PROBLEMS.

It may, in like manner, be proved that

$$\begin{vmatrix} a_1, & a_2, & a_3 \\ b_1, & b_2, & b_3 \\ c_1, & c_2, & c_3 \end{vmatrix} \begin{vmatrix} \alpha_1, & \alpha_2, & \alpha_3 \\ \beta_1, & \beta_2, & \beta_3 \\ \gamma_1, & \gamma_2, & \gamma_3 \end{vmatrix}$$

$$= \begin{vmatrix} \alpha_1 a_1 + \alpha_2 b_1 + \alpha_3 c_1, & \alpha_1 a_2 + \alpha_2 b_2 + \alpha_3 c_2, & \alpha_1 a_3 + \alpha_2 b_3 + \alpha_3 c_3 \\ \beta_1 a_1 + \beta_2 b_1 + \beta_3 c_1, & \beta_1 a_2 + \beta_2 b_2 + \beta_3 c_2, & \beta_1 a_3 + \beta_2 b_3 + \beta_3 c_3 \\ \gamma_1 a_1 + \gamma_2 b_1 + \gamma_3 c_1, & \gamma_1 a_2 + \gamma_2 b_2 + \gamma_3 c_2, & \gamma_1 a_3 + \gamma_2 b_3 + \gamma_3 c_3 \end{vmatrix}$$

and so on, for any number of rows and columns.

17. *If* (f, g, h), (f', g', h'), (f'', g'', h'') *be three points which form a conjugate triad with respect to the conic*

$$\phi(x, y, z) = ux^2 + vy^2 + wz^2 + 2u'yz + 2v'zx + 2w'xy = 0,$$

then

$$\begin{vmatrix} u, & w', & v' \\ w', & v, & u' \\ v', & u', & w \end{vmatrix} \begin{vmatrix} f, & g, & h \\ f', & g', & h' \\ f'', & g'', & h'' \end{vmatrix}^2$$

$$= \phi(f, g, h)\, \phi(f', g', h')\, \phi(f'', g'', h'').$$

For, generally,

$$\begin{vmatrix} u, & w', & v' \\ w', & v, & u' \\ v', & u', & w \end{vmatrix} \begin{vmatrix} f, & g, & h \\ f', & g', & h' \\ f'', & g'', & h'' \end{vmatrix}$$

$$= \begin{vmatrix} uf + w'g + v'h, & uf' + w'g' + v'h', & uf'' + w'g'' + v'h'' \\ w'f + vg + u'h, & w'f' + vg' + u'h', & w'f'' + vg'' + u'h'' \\ v'f + u'g + wh, & v'f' + u'g' + wh', & v'f'' + u'g'' + wh'' \end{vmatrix}$$

$$= \begin{vmatrix} \phi_f, & \phi_g, & \phi_h \\ \phi_{f'}, & \phi_{g'}, & \phi_{h'} \\ \phi_{f''}, & \phi_{g''}, & \phi_{h''} \end{vmatrix}$$

where ϕ_f is written for $\frac{1}{2}\frac{d\phi}{df}$, &c.

Hence
$$\begin{vmatrix} u, & w', & v' \\ w', & v, & u' \\ v', & u', & w \end{vmatrix} \begin{vmatrix} f, & g, & h \\ f', & g', & h' \\ f'', & g'', & h'' \end{vmatrix}^2$$

$$= \begin{vmatrix} f\phi_f + g\phi_g + h\phi_h, & f'\phi_f + g'\phi_g + h'\phi_h, & f''\phi_f + g''\phi_g + h''\phi_h \\ f\phi_{f'} + g\phi_{g'} + h\phi_{h'}, & f'\phi_{f'} + g'\phi_{g'} + h'\phi_{h'}, & f''\phi_{f'} + g''\phi_{g'} + h''\phi_{h'} \\ f\phi_{f''} + g\phi_{g''} + h\phi_{h''}, & f'\phi_{f''} + g'\phi_{g''} + h'\phi_{h''}, & f''\phi_{f''} + g''\phi_{g''} + h''\phi_{h''} \end{vmatrix}$$

But $f\phi_f + g\phi_g + h\phi_h = \phi(f, g, h)$, whatever f, g, h may be.

And $\quad\quad\quad f''\phi_f + g'\phi_g + h'\phi_h = 0.$

Similarly, all terms of similar form $= 0$.

Hence the theorem is proved.

18. *A triangle is inscribed in the conic*,

$$ux^2 + vy^2 + wz^2 + 2u'yz + 2v'zx + 2w'xy = 0,$$

two of its sides passing through the fixed points (f, g, h), (f', g', h'), *to find the envelope of the third.*

Call the fixed points K, K', and the angular points of the triangle PQR, RP passing through K, PQ through K'. Then, by projecting the conic into a circle and the line KK' to infinity, the lines RP, PQ will project into two lines always parallel to themselves, and therefore containing a constant angle, hence QR will project into a line always touching a circle concentric with the given one. Therefore, in the given problem, the envelope of QR will be a conic, having double contact with the given one along the line KK', and will therefore be represented by the equation

$$\lambda \phi(x, y, z) + \begin{vmatrix} f, & g, & h \\ f', & g', & h' \\ x, & y, & z \end{vmatrix}^2 = 0 \quad\quad\quad\ldots\ldots\ldots\ldots(1),$$

$\phi(x, y, z)$ being written, for shortness, instead of

$$ux^2 + vy^2 + wz^2 + 2u'yz + 2v'zx + 2w'xy,$$

and λ being a constant to be determined.

Now we observe, in the first place, that λ must be of two dimensions in f, g, h, of two in f', g', h', and of -1 in u, v, w, u', v', w'.

SUPPLEMENTARY PROBLEMS.   175

Next, let V be the point of intersection of two consecutive positions of QR. Then, if a triangle be inscribed in the conic so that two of its sides always pass through K', V, the envelope of the third side will pass through K. Hence (1) must be satisfied when we exchange x, y, z, with f, g, h.

Therefore we have

$$\lambda_{\prime}\phi(f, g, h) + \begin{vmatrix} x, & y, & z \\ f', & g', & h' \\ f, & g, & h \end{vmatrix} = 0,$$

$\lambda_{\prime}$ being what λ becomes when x, y, z are written for f, g, h. Hence

$$\lambda\phi(x, y, z) = \lambda_{\prime}\phi(f, g, h) \text{ identically,}$$

whence, λ involves $\phi(f, g, h)$ as a factor. Similarly it involves $\phi(f', g', h')$ as a factor. Hence we may write

$$\lambda = \frac{\phi(f, g, h)\,\phi(f', g', h')}{\mu},$$

μ being a function of u, v, w, u', v', w', of three dimensions, since λ is of -1.

The equation then becomes

$$\phi(f, g, h)\,\phi(f', g', h')\,\phi(\alpha, \beta, \gamma) + \mu \begin{vmatrix} f, & g, & h \\ f', & g', & h' \\ \alpha, & \beta, & \gamma \end{vmatrix}^2 = 0.$$

To determine μ, we may suppose, since it is independent of the co-ordinates of K, K', that each of these points lies on the polar of the other. Then, the envelope of QR must pass through the pole of KK', as may be seen by projecting KK' to infinity, for then QR will always pass through the centre of the conic. Hence, if (f'', g'', h'') be the polar of KK'

$$\phi(f, g, h)\,\phi(f', g', h')\,\phi(f'', g'', h'') + \mu \begin{vmatrix} f, & g, & h \\ f', & g', & h' \\ f'', & g'', & h'' \end{vmatrix}^2 = 0,$$

whence by Art. 17, $\mu = -\begin{vmatrix} u, & w', & v' \\ w', & v, & u' \\ v', & u', & w \end{vmatrix}.$

Therefore the required envelope is

$$\phi(f, g, h)\phi(f', g', h')\phi(x, y, z) = \begin{vmatrix} u, & w', & v' \\ w', & v, & u' \\ v', & u', & w \end{vmatrix} \begin{vmatrix} f, & g, & h \\ f', & g', & h' \\ x, & y, & z \end{vmatrix}^2.$$

19. *A triangle is described about a conic,* $x^2 + y^2 + z^2 = 0$, *two of its vertices move on fixed straight lines,* $lx + my + nz = 0$, $l'x + m'y + n'z = 0$; *to prove that the locus of the third vertex will be given by the equation*

$$(ll' + mm' + nn')^2 (x^2 + y^2 + z^2) + \begin{vmatrix} l, & m, & n \\ l', & m', & n' \\ x, & y, & z \end{vmatrix}^2 = 0.$$

It may be shewn, by reciprocating the theorem in the last article, that the locus will have double contact with the given conic along the pole of the intersection of the two given straight lines; hence its equation will be of the form

$$\lambda(x^2 + y^2 + z^2) + \begin{vmatrix} l, & m, & n \\ l', & m', & n' \\ x, & y, & z \end{vmatrix}^2 = 0 \ldots\ldots\ldots(1),$$

and it remains to determine λ.

For this purpose let the straight line $lx + my + nz = 0$ cut the given conic in P, P'. Let T be the pole of PP'.

Now, suppose one side of the triangle to become the tangent at P, then the other tangent through P will coincide with it, hence the required locus passes through the point of intersection of $l'x + m'y + n'z = 0$, with the tangent at P, and also with the tangent at P'.

Now, the co-ordinates of T are l, m, n, hence these two tangents are represented by the equation

$$(l^2 + m^2 + n^2)(x^2 + y^2 + z^2) - (lx + my + nz)^2 = 0 \ldots\ldots(2).$$

Hence (1) must be satisfied by the values of x, y, z, which satisfy (2), and also make

$$l'x + m'y + n'z = 0.$$

Now
$$\begin{vmatrix} l, & m, & n \\ l', & m', & n' \\ x, & y, & z \end{vmatrix}^2$$

TRILINEAR CO-ORDINATES OF FOCI.

$$= \begin{vmatrix} l^2+m^2+n^2, & ll'+mm'+nn', & lx+my+nz \\ ll'+mm'+nn', & l'^2+m'^2+n'^2, & l'x+m'y+n'z \\ lx+my+nz, & l'x+m'y+n'z, & x^2+y^2+z^2 \end{vmatrix}$$

which if $l'x + m'y + n'z = 0$, becomes

$$\begin{vmatrix} l^2+m^2+n^2, & ll'+mm'+nn', & lx+my+nz \\ ll'+mm'+nn', & l'^2+m'^2+n'^2, & 0 \\ lx+my+nz, & 0, & x^2+y^2+z^2 \end{vmatrix}$$

$$= (l^2+m^2+n^2)(l'^2+m'^2+n'^2)(x^2+y^2+z^2)$$
$$\qquad - (ll'+mm'+nn')^2(x^2+y^2+z^2)$$
$$\qquad - (l'^2+m'^2+n'^2)(lx+my+nz)^2$$
$$= -(ll'+mm'+nn')^2(x^2+y^2+z^2), \text{ if (2) be satisfied.}$$

Hence, by (1) $\lambda - (ll' + mm' + nn')^2 = 0$, identically, therefore, the equation of the required locus becomes

$$(ll'+mm'+nn')^2(x^2+y^2+z^2) + \begin{vmatrix} l, & m, & n \\ l', & m', & n' \\ x, & y, & z \end{vmatrix}^2 = 0.$$

TRILINEAR CO-ORDINATES OF THE FOCI OF A CONIC.

20. The following investigation of the trilinear co-ordinates of the foci does not introduce the conception of the imaginary circular points at infinity, or of imaginary tangents.

The trilinear co-ordinates of the focus of the conic

$$u\alpha^2 + v\beta^2 + w\gamma^2 + 2u'\beta\gamma + 2v'\gamma\alpha + 2w'\alpha\beta = 0,$$

may be investigated in the following manner. Draw two tangents to the conic parallel to $\alpha = 0$, and let f_1, f_2 be their respective distances from that line. Then, if f, g, h be the co-ordinates of a focus, we have

$$(f - f_1)(f_2 - f) = \text{the square on the semi-axis minor.}$$

F.   12

TRILINEAR CO-ORDINATES.

If the equation of either tangent be

$$\alpha'(b\beta + c\gamma) = (2\Delta - a\alpha')\alpha,$$

which represents a line parallel to, and at a distance α' from $\alpha = 0$, the two values of α' obtained by introducing the condition of tangency, will be f_1, f_2. Now, the condition of tangency is

$$\begin{vmatrix} 0, & a'\alpha - 2\Delta, & b\alpha', & c\alpha' \\ a\alpha' - 2\Delta, & u, & w', & v' \\ b\alpha', & w', & v, & u' \\ c\alpha', & v', & u', & w \end{vmatrix} = 0,$$

or
$$U(a\alpha' - 2\Delta)^2 + V(b\alpha')^2 + W(c\alpha')^2$$
$$+ 2U'b\alpha' \cdot c\alpha' + 2V'c\alpha'(a\alpha' - 2\Delta) + 2W'(a\alpha' - 2\Delta)b\alpha' = 0,$$

which may be written

$$(Ua^2 + Vb^2 + Wc^2 + 2U'bc + 2V'ca + 2W'ab)\alpha'^2$$
$$- 4\Delta(Ua + W'b + V'c)\alpha' + 4\Delta^2 U = 0.$$

Hence, the left-hand member of this equation is *identically* equal to

$$(Ua^2 + Vb^2 + Wc^2 + 2V'bc + 2V'ca + 2W'ab)(\alpha' - f_1)(\alpha' - f_2),$$

and therefore the square on the semi-axis minor

$$= -f^2 + 4\Delta \frac{(Ua + W'b + V'c)f - \Delta U}{Ua^2 + Vb^2 + Wc^2 + 2U'bc + 2V'ca + 2W'ab}.$$

Two similar expressions being obtained, we get for the determination of the foci, the equations

$$(Ua^2 + Vb^2 + Wc^2 + 2U'bc + 2V'ca + 2W'ab)f^2$$
$$- 4\Delta(Ua + W'b + V'c)f + 4\Delta^2 U$$
$$= (Ua^2 + Vb^2 + Wc^2 + 2U'bc + 2V'ca + 2W'ab)g^2$$
$$- 4\Delta(Vb + U'c + W'a)g + 4\Delta^2 V$$
$$= (Ua^2 + Vb^2 + Wc^2 + 2U'bc + 2V'ca + 2W'ab)h^2$$
$$- 4\Delta(Wc + V'a + U'b)h + 4\Delta^2 W$$

the same as those obtained in Chap. VI. Art. 33.

MISCELLANEOUS EXAMPLES.

1. Prove that the centre of the conic

$$\frac{1}{a\alpha} + \frac{1}{b\beta} + \frac{1}{c\gamma} = 0$$

coincides with the centre of gravity of the triangle of reference.

2. Prove that $\begin{vmatrix} 0, & 1, & 1, & 1 \\ 1, & 0, & z^2, & y^2 \\ 1, & z^2, & 0, & x^2 \\ 1, & y^2, & x^2, & 0 \end{vmatrix} = (x+y+z)(x-y-z)(y-z-x)(z-x-y).$

3. Prove that the square on the radius of the circle, described about the triangle of which the angular points are a, b, c, is

$$-\frac{1}{2} \frac{\begin{vmatrix} 0, & ab^2, & ac^2 \\ ba^2, & 0, & bc^2 \\ ca^2, & cb^2, & 0 \end{vmatrix}}{\begin{vmatrix} 0, & 1, & 1, & 1 \\ 1, & 0, & ab^2, & ac^2 \\ 1, & ba^2, & 0, & bc^2 \\ 1, & ca^2, & cb^2, & 0 \end{vmatrix}}$$

Investigate a similar expression for the square on the radius of the sphere, described about the tetrahedron of which the angular points are a, b, c, d.

4. S is a focus of a conic, PQ a chord subtending a constant angle at S; SR, ST are drawn meeting the tangents at P and Q in R, T respectively, so that the angles PSR, QST are constant; prove that RT always touches a conic having S for a focus, and a directrix in common with the given conic.

5. Prove that, if the conic $(l\alpha)^{\frac{1}{2}} + (m\beta)^{\frac{1}{2}} + (n\gamma)^{\frac{1}{2}} = 0$ be a parabola, its focus and directrix are given by the equations

$$\frac{l\alpha}{a^2} = \frac{m\beta}{b^2} = \frac{n\gamma}{c^2},$$

$$\frac{l\alpha}{\tan A} + \frac{m\beta}{\tan B} + \frac{n\gamma}{\tan C} = 0.$$

Hence prove that, if a parabola touch three straight lines, its directrix always passes through a fixed point. State, in geometrical language, the position of this point relatively to the three straight lines.

6. A system of parabolas is described so that a given triangle is self-conjugate with respect to each curve of the system; prove that the locus of the focus is a circle, that the directrix always passes through the centre of the circle described about the triangle, and that every parabola of the system touches the three straight lines which bisect each pair of sides of the triangle.

7. If P be any point on the circumference of a circle, O any fixed point, prove that the locus of the point, in which the tangent at P intersects the line which bisects OP at right angles, is a straight line.

8. A rectangular hyperbola circumscribes a triangle; shew that the loci of the poles of its sides are three straight lines forming another triangle, whose angular points lie on the sides of the first, where they are met by perpendiculars from the opposite angular points.

9. If ABC, $A'B'C'$ be two triangles, each of which is self-conjugate with regard to the same given conic, shew that another conic can be described about both.

10. If α, β, γ, δ be the distances of a point from four given straight lines, so connected that $l\alpha + m\beta + n\gamma + p\delta = 0$, prove that, if a conic be described, touching these four straight lines, the locus of either of its foci will be the curve of the third degree represented by the equation

$$\frac{l}{\alpha} + \frac{m}{\beta} + \frac{n}{\gamma} + \frac{p}{\delta} = 0.$$

EXAMPLES.

11. Prove that the polar reciprocal of a rectangular hyperbola with respect to any point S, is a conic, the sum of the squares on the semi-axes of which is equal to the square on the distance of its centre from S.

12. Two given conics are so related that each of their common tangents subtends a right angle at a given point. Prove that, if any two points be taken, one on each conic, so that the line joining them also subtends a right angle at that point, the envelope of this line will be a conic, of which that point is a focus.

13. In Example 2, p. 116, prove that if any conic (A) be drawn touching the directrices of the four conics, the polar of the given point with respect to it will be a tangent to a conic, having the given point as focus and touching the sides of the triangle; and that the tangents from the given point to A are at right angles to each other.

14. If, through a fixed point O, a straight line be drawn cutting the sides AB, AC of a triangle ABC in P, Q respectively, and BQ, CP be joined, prove that the locus of their point of intersection is a conic circumscribing the triangle ABC.

15. If ρ_a, ρ_b, ρ_c be the semi-diameters of a conic, respectively parallel to the sides of the triangle of reference, prove that the area of the conic is

$$8\pi \frac{\rho_a^2 \rho_b^2 \rho_c^2}{\sin A \sin B \sin C} \Sigma \left(\Sigma - \frac{\sin A}{\rho_a}\right)\left(\Sigma - \frac{\sin B}{\rho_b}\right)\left(\Sigma - \frac{\sin C}{\rho_c}\right)$$

where
$$2\Sigma = \frac{\sin A}{\rho_a} + \frac{\sin B}{\rho_b} + \frac{\sin C}{\rho_c}.$$

16. PQ is the chord of a conic, having its pole on the chord AB or AB produced; Qq is drawn parallel to AB meeting the conic in q; shew that Pq bisects the chord AB.

17. Similar circular arcs are described on the sides of a triangle ABC, their convexities being towards the interior of the triangle; shew that the locus of the radical centre of the three circles is the rectangular hyperbola

$$\frac{\sin(B-C)}{\alpha} + \frac{\sin(C-A)}{\beta} + \frac{\sin(A-B)}{\gamma} = 0.$$

18. Prove that, if r be either semi-axis of the curve represented by the equation

$$u\alpha^2 + v\beta^2 + w\gamma^2 + 2u'\beta\gamma + 2v'\gamma\alpha + 2w'\alpha\beta = 0,$$

the values of r will be the roots of the equation

$$\frac{a^2}{\left\{u + \frac{a}{bc}(a'u - bv' - cw')\right\}r^2 - as\cos A} + \frac{b^2}{\left\{v + \frac{b}{ca}(bv' - cw' - au')\right\}r^2 - bs\cos B}$$

$$+ \frac{c^2}{\left\{w + \frac{c}{ab}(cw' - au' - bv')\right\}r^2 - cs\cos C} = 0,$$

where
$$s = \frac{abc \begin{vmatrix} u, & w', & v' \\ w', & v, & u' \\ v', & u', & w \end{vmatrix}}{\begin{vmatrix} u, & w', & v', & a \\ w', & v, & u', & b \\ v', & u', & w, & c \\ a, & b, & c, & 0 \end{vmatrix}}.$$

19. If a triangle is self-conjugate with respect to each of a series of parabolas, the lines joining the middle points of its sides will be tangents: all the directrices will pass through O, the centre of the circumscribing circle: and the focal chords, which are the polars of O, will envelope an ellipse inscribed in the given triangle which has the nine-point circle for its auxiliary circle.

20. A conic circumscribes a triangle ABC, the tangents at the angular points meeting the opposite sides on the straight line DEF. The lines joining any point P on DEF to A, B, and C meet the conic again in A', B', C'; shew that the triangle ABC envelopes a fixed conic inscribed in ABC, and having double contact with the given conic at the points where they are met by DEF. Also the tangents at A', B', C' to the original conic meet $B'C'$, $C'A'$, $A'B'$ on points lying on DEF.

21. The anharmonic ratio of the pencil formed by joining a point on one of two conics to their four points of intersection is equal to the anharmonic range formed on a tangent to the other by their four common tangents.

EXAMPLES. 183

22. If the two pairs of straight lines represented by the equations $ax^2 + 2hxy + by^2 = 0$, $a'x^2 + 2h'xy + b'y^2 = 0$, form a harmonic pencil, the two straight lines of each pair not being conjugate, prove that $(ab' + ba' - 2hh')^2 = 36 (ab - h^2)(a'b' - h'^2)$.

23. The four common tangents to two conics intersect two and two on the sides of their common conjugate triad.

24. The locus of the centres of conics inscribed in a triangle and such that the centres of the escribed circles form a self-conjugate triad with respect to them is a straight line parallel to $a\alpha + b\beta + c\gamma = 0$ in areal co-ordinates.

25. A triangle ABC, right-angled at A, is inscribed in a rectangular hyperbola: tangents at B and C meet in P: prove that AB, AP, AC and the tangent at A form a harmonic pencil.

26. AB, CD are two fixed chords of a conic. A straight line APQ meets CD in P and the curve in Q, and on CQ a point R is taken so that PR subtends a constant angle at B: the locus of R will be a conic passing through B and C.

27. Conics circumscribing a triangle have a common tangent at the vertex; through this point a straight line is drawn: shew that the tangents at the various points where it cuts the curves all intersect on the base.

28. One conic touches OA, OB in A and B, and a second conic touches OB, OC in B and C: prove that the other common tangents to the two conics intersect on AC.

29. With any one of four given points as centre, a conic is described, self-conjugate with regard to the other three; prove that its asymptotes are parallel to the axes of the two parabolas which pass through the four given points.

30. A rectangular hyperbola passes through the angular points, and a parabola touches the sides, of a given triangle; shew that the tangents drawn to the parabola, from one of the points where the hyperbola cuts the directrix of the parabola, are parallel to the asymptotes of the hyperbola. Which of the two points on the directrix is to be taken? When they coincide, shew that either curve is the polar reciprocal of the other with respect to the coincident points.

31. The triangular co-ordinates of the two circular points at infinity are given by the equations

$$-\frac{x}{a} = \frac{y}{b\epsilon^{\pm\sqrt{-1}C}} = \frac{z}{c\epsilon^{\mp\sqrt{-1}B}}.$$

32. If each angular point of a triangle be joined with the points in which its opposite side is cut by a given conic, the six straight lines thus formed will be tangents to another conic.

33. With each of four given straight lines as directrix, two conics are described, self-conjugate with regard to the other three; prove that the four pairs of conics thus obtained, will have the same pair of points as foci corresponding to the given directrix.

34. If a triangle be self-conjugate to a rectangular hyperbola and any conic be described, touching the sides of the triangle, each focus of this conic will lie on the polar of the other with respect to the rectangular hyperbola.

35. If abc, def, be two triangles in different planes, prove that the product of their areas into the cosine of the inclination of their planes is

$$-\frac{1}{16} \begin{vmatrix} 0, & 1, & 1, & 1 \\ 1, & ad^2, & ae^2, & af^2, \\ 1, & bd^2, & be^2, & bf^2, \\ 1, & cd^2, & ce^2, & cf^2, \end{vmatrix}$$

36. A triangle is described about one conic section and circumscribed about another. Prove that a conic section may be described, touching one side of the triangle, and the opposite tangent to the outer conic section, and passing through the four points of intersection of the two conic sections.

37. Prove that the self-conjugate circles of the triangles formed by each three of four given straight lines have a common radical axis.

www.ingramcontent.com/pod-product-compliance
Lightning Source LLC
Chambersburg PA
CBHW032128160426
43197CB00008B/563